Stretch Blow Molding

PLASTICS DESIGN LIBRARY (PDL)
PDL HANDBOOK SERIES

Series Editor: Sina Ebnesajjad, PhD (sina@FluoroConsultants.com)
President, FluoroConsultants Group, LLC
Chadds Ford, PA, USA
www.FluoroConsultants.com

The **PDL Handbook Series** is aimed at a wide range of engineers and other professionals working in the plastics industry, and related sectors using plastics and adhesives.

PDL is a series of data books, reference works and practical guides covering plastics engineering, applications, processing, and manufacturing, and applied aspects of polymer science, elastomers and adhesives.

Recent titles in the series

Biopolymers: Processing and Products, Michael Niaounakis (ISBN: 9780323266987)
Biopolymers: Reuse, Recycling, and Disposal, Michael Niaounakis (ISBN: 9781455731459)
Carbon Nanotube Reinforced Composites, Marcio Loos (ISBN: 9781455731954)
Extrusion, 2e, John Wagner & Eldridge Mount (ISBN: 9781437734812)
Fluoroplastics, Volume 1, 2e, Sina Ebnesajjad (ISBN: 9781455731992)
Handbook of Biopolymers and Biodegradable Plastics, Sina Ebnesajjad (ISBN: 9781455728343)
Handbook of Molded Part Shrinkage and Warpage, Jerry Fischer (ISBN: 9781455725977)
Handbook of Polymer Applications in Medicine and Medical Devices, Kayvon Modjarrad & Sina Ebnesajjad (ISBN: 9780323228053)
Handbook of Thermoplastic Elastomers, Jiri G Drobny (ISBN: 9780323221368)
Handbook of Thermoset Plastics, 2e, Hanna Dodiuk & Sidney Goodman (ISBN: 9781455731077)
High Performance Polymers, 2e, Johannes Karl Fink (ISBN: 9780323312226)
Introduction to Fluoropolymers, Sina Ebnesajjad (ISBN: 9781455774425)
Ionizing Radiation and Polymers, Jiri G Drobny (ISBN: 9781455778812)
Manufacturing Flexible Packaging, Thomas Dunn (ISBN: 9780323264365)
Plastic Films in Food Packaging, Sina Ebnesajjad (ISBN: 9781455731121)
Plastics in Medical Devices, 2e, Vinny Sastri (ISBN: 9781455732012)
Polylactic Acid, Rahmat et al. (ISBN: 9781437744590)
Polyvinyl Fluoride, Sina Ebnesajjad (ISBN: 9781455778850)
Reactive Polymers, 2e, Johannes Karl Fink (ISBN: 9781455731497)
The Effect of Creep and Other Time Related Factors on Plastics and Elastomers, 3e, Laurence McKeen (ISBN: 9780323353137)
The Effect of Long Term Thermal Exposure on Plastics and Elastomers, Laurence McKeen (ISBN: 9780323221085)
The Effect of Sterilization on Plastics and Elastomers, 3e, Laurence McKeen (ISBN: 9781455725984)
The Effect of Temperature and Other Factors on Plastics and Elastomers, 3e, Laurence McKeen (ISBN: 9780323310161)
The Effect of UV Light and Weather on Plastics and Elastomers, 3e, Laurence McKeen (ISBN: 9781455728510)
Thermoforming of Single and Multilayer Laminates, Ali Ashter (ISBN: 9781455731725)
Thermoplastics and Thermoplastic Composites, 2e, Michel Biron (ISBN: 9781455778980)
Thermosets and Composites, 2e, Michel Biron (ISBN: 9781455731244)

To submit a new book proposal for the series, or place an order, please contact David Jackson, Acquisitions Editor
david.jackson@elsevier.com

Stretch Blow Molding

Third edition

O. Brandau
Member of Society of Plastics Engineers
Member of Mensa Canada

ELSEVIER

AMSTERDAM • BOSTON • HEIDELBERG • LONDON
NEW YORK • OXFORD • PARIS • SAN DIEGO
SAN FRANCISCO • SINGAPORE • SYDNEY • TOKYO
William Andrew is an imprint of Elsevier

William Andrew is an imprint of Elsevier
The Boulevard, Langford Lane, Kidlington, Oxford, OX5 1GB, United Kingdom
50 Hampshire Street, 5th Floor, Cambridge, MA 02139, United States

Library of Congress Cataloging-in-Publication Data
A catalog record for this book is available from the Library of Congress

British Library Cataloguing-in-Publication Data
A catalogue record for this book is available from the British Library

ISBN: 978-0-323-46177-1

Publisher: Matthew Deans
Acquisition Editor: David Jackson
Editorial Project Manager: Edward Payne
Production Project Manager: Lisa Jones
Designer: Greg Harris

Typeset by Thomson Digital

Contents

Preface to the Third Edition

It has been shown that all businesses and technologies follow a certain progress over time. In the beginning R&D costs lead to losses until the first products can be sold. A period of rapid growth follows and market leaders emerge. In the case of PET these were certainly Husky Injection Molding Systems for injection and Sidel for blow molding machines. During the later part of this expansion acquisitions tend to take place. Both aforementioned companies were sold, and are no longer owned by the original entrepreneurs. As products reach maturity and/or the market has been saturated the inevitable decline follows. Competitors catch up with technology that cannot be improved indefinitely and rapid growth is no longer sustainable. Water bottle sales offer a good example of this. We saw growth rates of 10% and higher year over year in the 1990s but this slowed down to a mere 3–5%. At the same time water bottle weights have decreased from 18 to 8 g and will probably hover in this area as there are limits on how light one can make them. The technology lead early adaptors had has shrunk considerably and competitors from all over the world have caught up to a large degree.

It seems we are at the edge of the growth area and it will depend on how many more innovations the market leaders can produce whether the industry as a whole will go into decline. This does not mean that blow machines will no longer be built but rather that the growth will go down to 2% or 3% with some years ending up in negative territory. Also, market leaders' market share will shrink as competitors with lower overhead will be able to compete effectively. A good example how this may play out can be seen in extrusion blow molding, an industry I grew up in. Bekum out of Germany was the undisputed leader in shuttle machine technology up to the early 1980s. Over the last few years they have closed three manufacturing sites of the original five.

While carbonated soft drinks sales are in decline and water has faced criticism over waste issues conversion from glass, carton, and other plastics could be the future growth area. PE and PET are now at par pricewise with prices for both resins coming down remarkably and PET has many advantages over PE and other plastics. Its recyclability is unmatched especially versus carton and it is in the interest of the industry to push for more recycling. This will reduce its carbon footprint and relieve pressure from environmentalists giving PET its well-deserved green shine!

When it comes to blow molding machinery we are reaching the end of improvements in speed. The fastest machines now produce 2400 bottles/cavity per hour. At this speed cycle time translates into a mere 1.5 s and this includes all the mechanical functions of the machine leaving maybe 0.75 s for the process. There are improvements in heating and handling that show great promise (Chapter 4) and the tendency will go toward defect elimination, automatic machine adjustments, and cost savings.

In short, we are moving into a less exiting time when the winners will be the ones that can make the most cost-effective, high-quality machines, and subsequently bottles.

O. Brandau
Feb. 2016

Acknowledgments

The following individuals and companies have contributed to this book in one way or another and the author wishes to express his gratitude to:

Laura Martin

Tony Padget

Rick Unterlander

Accuform, Czech Republic	www.b-sim.com
AEC, Inc, USA	www.aecinternet.com
Adphos Group, USA	www.adphos.com
AGR International Inc, USA	www.agrintl.com
Amcor PET Packaging, USA	www.amcor.com.au/
Amsler Equipment Inc., Canada	www.amslerequipment.net
Atelier Francois, Belgium	www.afcompressors.com
Belvac Production Machinery, USA	www.belvac.com
Bekum America Corporation, USA	www.bekumamerica.com
Bericap, Germany	www.bericap.com
Chumpower, Taiwan	www.chumpower.com
Colormatrix, USA	www.colormatrix.com
Compound Metal Coatings, Canada	www.cmcnickelplating.com/cms
Concordia Development S.r.L., Italy	www.ziggypack.net
Corvaglia Mould AG, Switzerland	www.corvaglia.ch
CYPET Technologies, Cyprus	www.cypet.eu
Garrtech Inc, Canada	www.garrtech.com
Goldco Industries, Inc., USA	www.goldcointernational.com
Graham Packaging, USA	www.grahampackaging.com
Hallink Molds Ltd, Canada	www.hallink.com
Ingersoll-Rand Company Limited, USA	www.air.irco.com
Intravis Vision Systems, Germany	www.intravis.com
Invista, USA	www.invista.com
KHS-Corpoplast, Germany	www.khscorpoplast.com
Kortec Inc., USA	www.kortec.com
Krones AG, Germany	www.krones.de
Lanfranchi S.r.l., Italy	www.lanfranchi.it
Mayr Corporation, Germany	www.mayrcorp.com
Mega Machinery, China	www.megamachinery.com
Nissei ASB, Japan	www.nisseiasb.co.jp

Philips, Netherlands	www.infrared.philips.com
Posimat, Spain	www.posimat.com
Pressco Technology, Inc	www.pressco.com
Riondé S.A, France	www.rionde-sa.fr
Schaub Chiller Service, USA	www.chillers.com
Siapi, Italy	www.siapi.it
Sidel, France	www.sidel.com
SIPA, Italy	www.sipa.it
Synventive, USA	www.synventive.com
UAB Terekas, Lithuania	www.terekas.lt

Introduction

Reheat stretch blow molding (RSBM) is part of the two-stage process of making bottles from polyethylene terephthalate (PET) or other resins. During the first stage injection machines produce vial or test-tube shaped "preforms." The necks of preforms are fully finished but the diameter and length is much smaller than the bottle into which it will be transformed during the RSBM process. During this transformation the material undergoes significant changes in molecular orientation making PET bottles virtually unbreakable, lightweight, and enhancing various barrier properties while keeping the clarity that is also present in the preforms.

PET bottle production has enjoyed tremendous success over the past 30 years. In 2006, 12.3 million tons of PET resin worldwide was converted into containers and PET still enjoys the highest growth rate of any major plastic, although this rate has slowed from a stunning 20% in 1990 to a more moderate 5–6% today.

There are three different ways of making a PET bottle: the single-stage, integrated two-stage, and two-stage process This book concentrates on the latter but I have added a chapter on the single-stage process giving this important process its due. There are several advantages of the two-stage manufacturing model in comparison with the other two. For one thing, injection and blow molding are completely independent of each other and can therefore be optimized separately. It also means that preforms may be stored, shipped across great distances, even countries, and used when required. In our globalized world this has helped spread the process as more and more different preforms become available.

While RSBM changed from a niche application to a very important plastic process, there is an astounding lack of published material on the subject. Magazines and conferences are the means of information flow, while books relegate RSBM to a few chapters. This book attempts to give the industry its due. It is written with the people in mind who produce millions of PET containers every day. It bridges the gap between a purely theoretical work and the operational part of a stretch blow molding machine manual.

While it cannot detail the layout of features and controls of any specific machine, the reader will find all relevant *process* data that he or she may need to make informed decisions both at the desk and on the shop floor. The book is written for both novices and experts. Novices may follow the structure of the book and expand their knowledge of these steps. Experts might find interesting details, even in the more general sections, while benefiting most from some of the in-depth descriptions.

The reader may browse through the different sections at random but Chapters 3 and 7 contain basic information that is needed for a full understanding of Chapters 8 and 13. Chapter 4 gives a more general view of the machinery used while processors and engineers will find more detailed information in Chapter 5.

Companies using the RSBM process are very protective of their accumulated in-house expertise, rendering PET processing something of a mystery. This book attempts to make relevant information accessible to a broad audience. Enabling more people to produce high-quality PET bottles will benefit the industry as a whole and encourage development of new applications.

1 Short History of Stretch Blow Molding

The idea of reheating a thermoplastic material and then stretching it to enhance its properties was first employed in extruded sheet in the 1930s. But it took until the 1970s for Nathaniel Wyeth and his staff to blow the first polyethylene terephthalate (PET) bottle from an injection-molded PET perform at the DuPont company. At the same time Bekum Maschinenfabriken in Germany had commercialized a similar process, stretch blow molding an extrusion blow molded PVC perform in what we would call today a single-stage process. Oriented PVC has similar oxygen and water barrier, and even carbonation retention, than PET. Bekum's Oriented PVC machines featured a double carriage where one side blew a preform from an extruded parison that was then transferred to the other side where the bottle was stretched and blown. This yielded a lightweight bottle with superior properties and was successfully used to produce a variety of containers. However, PVC became environmentally suspect and PET is not suited to a process that requires what extrusion blow molders call "hang strength," the ability of the material to sustain shape at melt temperature against gravity. Another problem with the PVC process was its inability to be scaled up easily (Fig. 1.1).

Meanwhile several US-based companies had developed machinery to produce stretch-blown PET bottles. Cincinnati Milacron's RHB-5 machine reheated performs neck side up in four lanes, then stretching and blowing them in a four-cavity mold. All molds moved at the same time and machines of this type are referred to as linear or in-line machines. Initially, output was limited to 2800 bottles per hour (bph) but later versions boosted output to 4000 bph before Cincinnati stopped producing them in the early1990s.

Meanwhile in Europe the German company Gildameister (later to become Corpoplast and today KHS Corpoplast) and the French company Sidel were developing machines for PET production. Sidel had produced extrusion blow molding machines using horizontal wheels. In a wheel machine each individual mold cavity opens and closes in sequence and machines of this type are called rotary machines. In the late 1970s Sidel started experimenting with the use of this concept in the PET stretch blow molding process. By 1980s, Sidel had built the first prototype machine that would start an unparalleled success in the blow molding industry,

Stretch Blow Molding. http://dx.doi.org/10.1016/B978-0-323-46177-1.00001-9

Figure 1.1 Bekum's double-sided extrusion stretch blow machine for PVC where preforms are blown in the inner carriages and bottles in the outer ones. Picture courtesy of Bekum America Corporation.

propelling Sidel from a midsize machine manufacturer to a billion dollar company (Fig. 1.2).

Today companies such as Krones, Smiform, and SIPA have all developed rotary machines of their own and this competitive pressure has driven prices down, opening new applications for bottle blowing. Blow molding speeds have also driven costs down, while 1000 bottles/cavity per h was the benchmark for many years, today's machines feature outputs of up to 2200 bottles/cavity per h.

The first "killer application" for PET was the 2-L bottle for carbonated soft drinks (CSD), introduced in 1978. The first bottles featured a dome-shaped bottom ideally suited to sustain internal pressures that routinely reach 5 bar (70 psi). This required an additional plastic component, called a base cup, to be glued to the bottom in a secondary operation in order for the bottle to stand up. However, cost as well as recycling considerations (glue residue) encouraged the development of a one-piece bottle. The breakthrough came with the design of the so-called Petaloid base, a thick, mostly amorphous center disk surrounded by five blown feet. Granted as patent to the Continental Can Company in 1971, it caused controversy with three other patents and litigation ensued over several years. It took until the early 1990s before one-piece bottles came off the conveyors of

Figure 1.2 Rotary high-speed machines such as this blow molder produce the bulk of PET bottles. Photo courtesy of KHS Corpoplast.

reheat stretch blow molding machines and completely replaced two-piece bottles within a few short years.

By the mid-1990s, soft drink companies agreed to lower shelf life requirements and so opened the way for the extremely successful launch of 20 oz and 500 mL containers. At the time of writing in early 2011, it was water and a whole new line of beverages that did not even exist a few years ago that was the key drivers for PET growth. Hot-fill juices and so-called neutraceuticals have raised the demands imposed on today's PET bottles and the industry has responded with a wealth of new technologies. Recent developments aim to eliminate the unsightly vacuum panels needed for controlled shrinkage of the PET bottle during cooling of a hot-filled product. Multilayer preforms and coating technologies increase shelf life and so will open the way for even smaller CSD packages and the replacement of glass in a new set of applications. At the time it is unclear whether coatings or multilayer technologies will prevail as the preferred choice of packaging, so variety of methods will be called for, to meet an ever-increasing variety of packaging demands.

On the horizon we can see PET entering the retort arena, used for packages that need exposure to typically 125°C (257°F) for a number of minutes, and that are all filled in cans and glass today. The PET bottle's crystallinity levels will have to be substantially increased to allow the use of PET here. At this time the highest temperature PET is being exposed to commercially is 148°C (300°F) in a jar with a metal lid. Barrier enhancements will allow extended shelf life (ESL) milk to be packaged in PET among other goods that require a long shelf life. Improved ways of injecting preforms and blowing bottles will extend the industry ability to deliver a safe, environmentally sound, and economical package to consumers.

2 Material Basics

Chapter Outline

2.1 Manufacture and States of PET

PET belongs to the group of materials known as thermoplastic polymers. The application of heat causes the softening and deformation of thermoplastics. In contrast, thermosets cure or solidify with the application of heat, and simply burn with continued heating.

Like all polymers, PET is a large molecule consisting of chains of repeating units. The PET used for bottles typically has about 100–140 of the repeating units shown in Fig. 2.1.

A monomer is a single unit which is repeated to form a polymer chain (*Greek "mono"* one; "*meros*" part). Polymerization is the name given to the types of reactions where many monomer units are chemically linked to form polymers ("*polys*" many).

A resin with only one type of monomer is called a homopolymer. Copolymer resins are the result of modifying the homopolymer chain with varying amounts of a second monomer (or comonomer) to change

Stretch Blow Molding. http://dx.doi.org/10.1016/B978-0-323-46177-1.00002-0

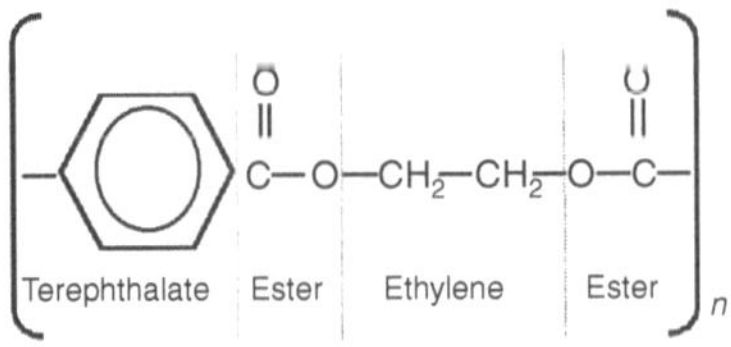

Figure 2.1 The ring structure makes PET tough while the ethylene component gives it flexibility.

some of the performance properties of the resin. This can be represented by:

Homopolymer	AAAAAAAAAAAAAAAAAAA
Copolymer	ABAAABAAAAABAAABBAA

PET is manufactured as a homopolymer or copolymer.

Manufacture of PET

There are a few chemical routes to manufacturing PET, but basically a compound with two acids, such as terephthalic acid (TPA), is esterified with a compound with two alcohols, ethylene glycol (EG). Since there are two functional groups on each component, they can continue to link up to form long chains. Water is a by-product of this process. This esterification reaction is reversible, and this is the key to understanding much of the behavior of PET (Fig. 2.2).

Commercially the polymerization is done in two stages. Melt phase condensation results in molten polymer with about 100 repeat units (IV, as explained later, is about 0.6). The melt is pelletized, and can be used for some applications such as fiber at this point.

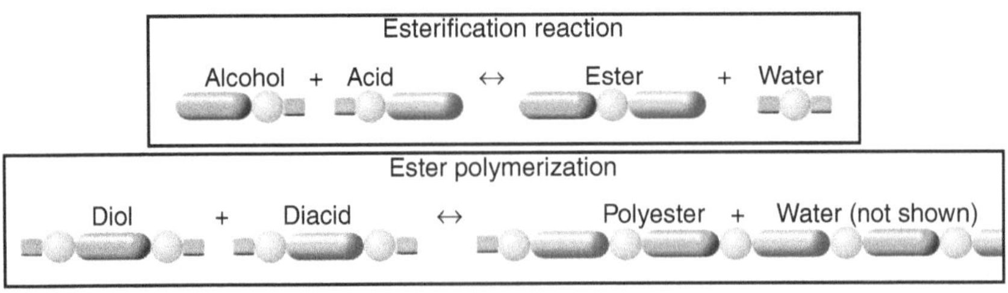

Figure 2.2 An alcohol and an acid form the ester groups of PET that make it a polyester.

To continue the polymerization, a process called "solid stating" is needed. Solid stating produces high molecular weight PET needed for fabricating bottles.

Catalysts

Different catalysts are required for the two main chemical routes to manufacture PET. Special catalyst combinations can be used to influence the side reactions, to reduce the amount of diethylene glycol (DEG) or acetaldehyde (AA), or to improve the color. Since the catalyst residues remain in the PET, they are still present during drying and processing. Therefore, different grades of PET from different manufacturers react differently if not processed at optimum conditions. For example, the DMT process (used chiefly by Eastman) requires an additional catalyst which may result in a greater tendency of the resin to oxidize or turn yellow when over-dried.

PET is a Linear Condensation Polymer

PET does not branch: each molecule is a long "linear" chain. In addition, because it is formed by a reversible condensation reaction, it has a very simple distribution of molecular weights, or chain lengths. The result as far as end-users are concerned is that the chemical structure of a grade of PET can be described quite completely by only two measures: intrinsic viscosity (IV), which is a measure of molecular weight, and the copolymer content. In contrast, a polymer such as polyethylene can have unique molecular weight distributions and widely varying degrees and types of branching, which affect processing and performance profoundly.

Intrinsic Viscosity

The properties of the PET polymer are largely dependent upon the average molecular weight, or the average number of repeat units of the polymer chains. This is usually determined by measurement of the intrinsic viscosity, or IV, as explained later. The relationship between molecular weight and IV is fairly linear.

High IV PET has a higher molecular weight than low IV PET. The longer chains not only give the resin better properties in the final product, but also affect the processing in predictable ways. The range of IVs used for PET bottle manufacturing is from about 0.73 to 0.86.

Copolymer Content

PET copolymers are made by replacing some percent of one of the starting components with a different monomer. Eastman uses cyclohexane dimethanol (CHDM) to replace part of the EG. Most other resin manufacturers use isophthalic acid (IPA), which is also called purified isophthalic acid (PIA), to replace part of the TPA. The copolymers therefore have structures such as:

PET	TETETE...
PET-*co*-CHDM	TETCTE...
PET-*co*-IPA	TEIETE...

DEG, a by-product of the polymerization reaction, is another comonomer which lowers the melt temperature but is not as effective at slowing down crystallization rates. DEG takes the place of EG in the chain.

Several advantages are gained by using the copolymer especially in preform molding applications:

1. Copolymers crystallize more slowly than homopolymers, making it easier to fabricate clear preforms (see Section 2.2).
2. Copolymers are easier to melt in the extruder as a result of the lower melting point and lower maximum degree of crystallinity.
3. Copolymers impart better stress-crack resistance to the bottle (see Embrittlement and Stress Cracking).

Some of the generalized effects of IV and copolymer content are outlined here.

Process/Performance Parameter	**Effect of Increased IV**	**Effect of Increased Copolymer Content**
Crystallization rate	⇓	⇓
Extruder motor load	⇑	⇓
Natural stretch ratio (NSR)	⇓	⇑
Orientation	⇑	⇓
Sidewall thickness	⇑	⇓
AA-generating potential	⇑	⇓
Aging rate	⇓	⇓
Stress-crack resistance		⇑

2.2 Crystallization of PET

PET is a semicrystalline resin. The word "crystalline" refers to a region of ordered chain arrangement, as opposed to "amorphous" where the polymer chains lack order. Melted PET, by definition, is amorphous.

When polymers are in an amorphous state, the molecular chains can be compared to a tangled web of spaghetti or springs. The analogy to tangled, stretched springs is particularly suitable for semicrystalline polymers because under certain conditions the polymer chains tend to coil into ordered structures, forming crystalline regions. The repeating units of the homopolymer chain fit together neatly, forming a close-packed array, which has a higher density than the amorphous state.

Density measurement is commonly used to determine the degree of crystallinity. At room temperature, amorphous PET has a density of $1.335 g/m^3$. The calculated density of perfect PET crystal is $1.455\ g/m^3$. The density of a semicrystalline sample with x fraction crystallinity is:

$$\frac{1}{\rho} = \frac{x}{1.455} + \frac{(1-x)}{1.335}$$

The crystal structure has a lower energy state than the amorphous arrangement, so it is the favored arrangement. Because polymer molecules are long and entangled, however, the amorphous state can be "frozen in" by rapidly cooling the PET melt. Crystallization can occur at any temperature at which the polymer chains have sufficient mobility to rearrange themselves. The rate of crystallization is a function of the temperature, the IV of the polymer, and any comonomer content. Polymers are very rarely able to crystallize completely.

The temperature range for crystallization is between the glass transition temperature (T_g) and the melt temperature (T_m). Below T_g, the resin is described as being "glassy" because the mobility of the polymer chains is greatly reduced, and are essentially locked in place regardless of whether they are in an amorphous or crystalline state. Above T_m, the polymer chains have too much energy to form stable ordered structures, and the molten resin is amorphous. Between T_g and T_m, the polymer chains have enough energy to rearrange themselves into the most thermodynamically favored structure, so the resin crystallizes (Fig. 2.3).

Thermally induced crystals are arranged in structures called "spherulites," since they start from a point source (nucleation site) and grow, radiating outward, in a spherical pattern. This must be avoided since the crystallized regions cause haze, destroying the clarity of the preform, and prevent proper stretch blow molding. The mold cooling becomes essential and determines the quality of the preform at this stage.

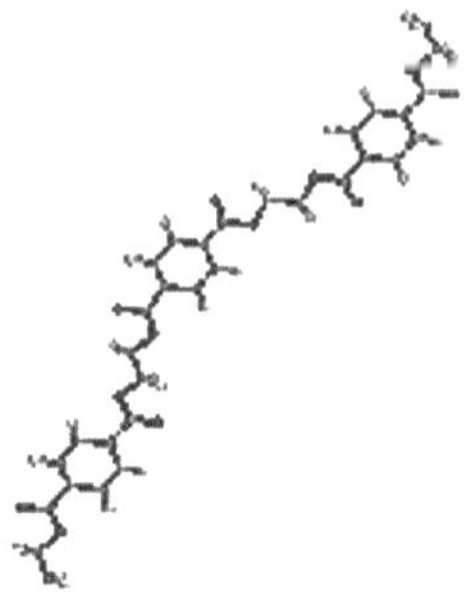

Figure 2.3 Rendering of a PET chain.

An increase in IV reduces the rate of crystallization by lengthening the polymer chains, making it more difficult for a given chain to disentangle itself from other chains and form an ordered crystal. Copolymer content changes the molecular structure of the chains, inhibiting crystallization by essentially introducing a unit into the chain which may not easily form crystals due to structural differences, or disrupts the crystalline pattern.

"Extended Chain" or "Oriented" Crystallization

During stretch blow molding, the amorphous chains in the preform are stretched and oriented, and a different form of crystallinity is developed. The chains are aligned in the direction of stress, orienting the chains and imposing a linear ordered structure throughout the area of applied stress. This "extended chain" or stress-induced crystallinity is necessary in the blow-molded container for mechanical strength.

Summary

PET occurs in three different states:

Amorphous, nonoriented, and clear, such as preforms and melted plastic resin
Thermally (by means of temperature) crystallized, such as resin pellets
Strain-induced crystallized, such as bottle sidewalls.

PET is transformed several times as it goes from pellet to preform to bottle.

- As resin pellets, PET is thermally crystallized to a level of 50–70%. Thermally induced crystals are arranged in large structures called spherulites which reflect light. Therefore, PET appears white.

- During the injection process, these crystals are melted, resulting in an amorphous melt, which is injected into the perform mold cavities. The preform is rapidly cooled down to avoid recrystallization. Preforms therefore do not have a crystal structure. This state is called amorphous. In the amorphous state the molecular chains show no orientation and no crystallinity, and their appearance has been compared with a bowl of spaghetti. There is nothing to reflect light and therefore the PET is clear. It also has little strength or barrier properties.
- In the reheat stretch blow machine the material is forced by the stretch rod and blow air to orient in the axial and hoop direction forming small, strain-induced crystals. These crystals do not reflect light and the bottle appears clear. It also has higher strength and barrier properties. Crystallization levels of up to 25% can be achieved in the bottle sidewall given the correct preform design and process conditions. (Higher crystallinity levels are achieved in the heat-set process, see Chapter 8, Section 8.3.)

The finished bottle will have amorphous portions in the neck and gate area where the bottle was not stretched, oriented portions in the sidewalls and sometimes thermally crystallized portions around the gate, a common preform defect that cannot be corrected during blow molding (Fig. 2.4).

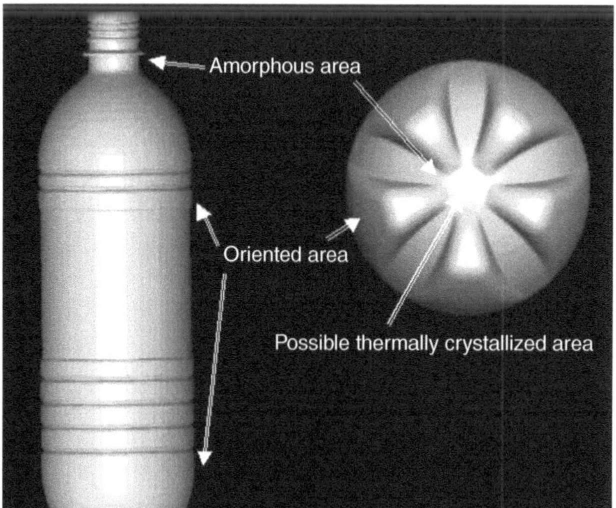

Figure 2.4 Different states of PET are present in each bottle.

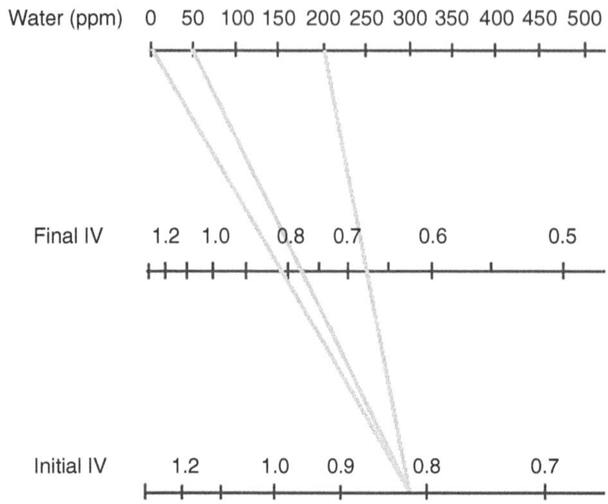

Figure 2.5 Three moisture levels, three resulting IV results. A material of 0.82 IV will be reduced to an IV of 0.68 when processed with a moisture content of 200 ppm.

2.3 Drying of PET

Because PET is hygroscopic it must be dried before it can be injected. The maximum amount of water that can be in the resin when it is in the extruder throat is 50 ppm. This residual moisture will react with the PET in the extruder and lead to an acceptable drop of 0.03–0.04 in IV. Higher moisture levels will lead to much higher IV drops, rendering the material unsuitable for the application (Fig. 2.5).

The correct drying parameters are a combination of time and temperature at a certain airflow. Modern dryers are able to generate the required air flow of 4 m^3/h per kg per h (1 cfm/lbs. per h). Under these conditions, processors must calculate or determine by practical experiment what the residence time of the resin in the hopper is for a given job. To do this practically, a handful of color pellets is placed on top of the resin in the hopper with the time noted. The colored pellets will eventually show up in the preforms and the time can then be measured. Depending on the position of the resin in the hopper, drying times differ with the resin in the center of the hopper traveling up to 20% faster. Therefore a median residence time must be chosen. Once this residence time has been established, the proper drying temperature can be chosen as from the graph to the left (Fig. 2.6).

Modern dryers now offer the ability to calculate residence time. This is achieved either by not filling the hopper to the top but a height where the

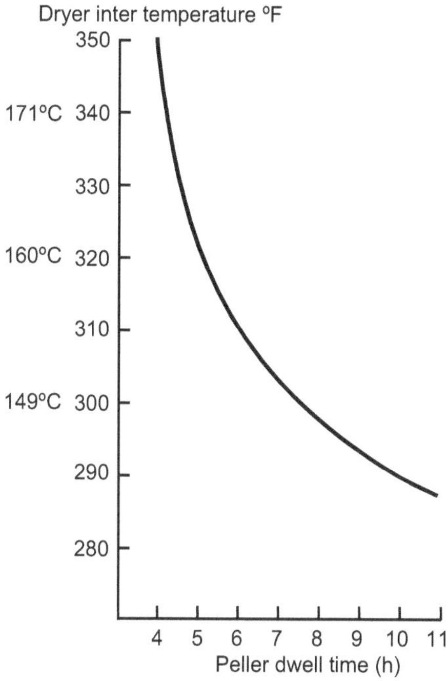

Figure 2.6 Drying time must be chosen to match the residence time of the resin in the hopper.

contents are commensurate with the chosen residence time. Or air flow is controlled via an inverter on the fan motor that reduces air flow when the dryer detects too long a residence time.

Maximum drying temperature is 180°C (356°F). While this temperature will lead to oxidation, which shows up as a yellowing of the resin, the reality is that the resin is not exposed to it the whole time during its passage through the dryer. The air that enters typically through the bottom, will gradually cool as it moves upward, which can easily be observed at the gauge for the air outlet. However, when drying at high temperatures it is paramount that a residence time of maximal 6 h is observed. That means that the dryer temperature must be reduced to a standby value of around 100°C (212°F) when there is a production interruption.

Improper drying and the resultant drop in IV change the inflation behavior of the preform in that, the preform will inflate under lower pressure because the NSR is greater. In turn, this will lead to less orientation and weaker bottles. Preform designers should know this connection in case problems arise during production, which are all too easily blamed on preform design.

2.4 Other Consequences of Insufficient Drying

Drying also plays a role in energy consumption. When considering how the material temperature is increased from room to melt temperature there are obviously three sources: the dryer, the heater bands at the barrel and hot runner, and the shear action of the screw. Dryers offer a very high (up to 98%) efficiency in heating as they are well insulated and heat the resin directly. The shear action and barrel heater bands by contrast heat the resin less efficiently (about 60%). For the processor this means that the higher the barrel inlet temperature is (up to the maximum indicated above), the less energy the process requires. There is therefore a strong argument to use the latest dryer technology that allows adjusting the residence time to output in order to have the most cost-effective operation.

When using all-electric screw drives running the dryer in this way will also reduce the load on the motor driving the screw. Overheating charge motors seem to be a recurrent problem with all-electric machines and proper drying will reduce this issue.

2.5 Behavior in the Injection Mold

We will not discuss the melting and visco-elastic flow of the material in the extruder barrel as they do not pertain as much to the preform design. But the injection part is important for designers to understand because of the particular opportunities and process limits as well as possible defects that will then affect the blown bottles (Fig. 2.7).

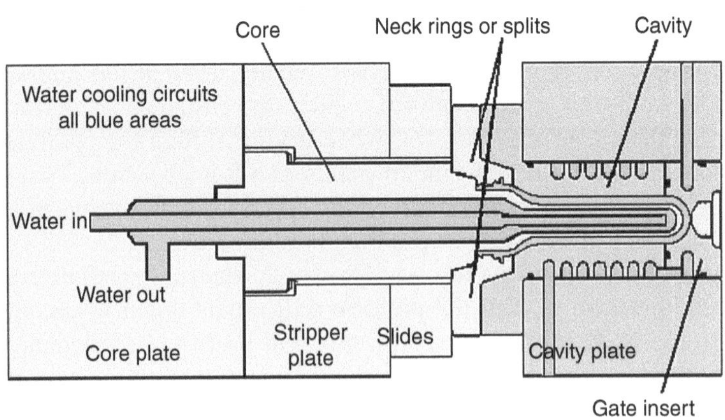

Figure 2.7 The various components of a typical injection mold.

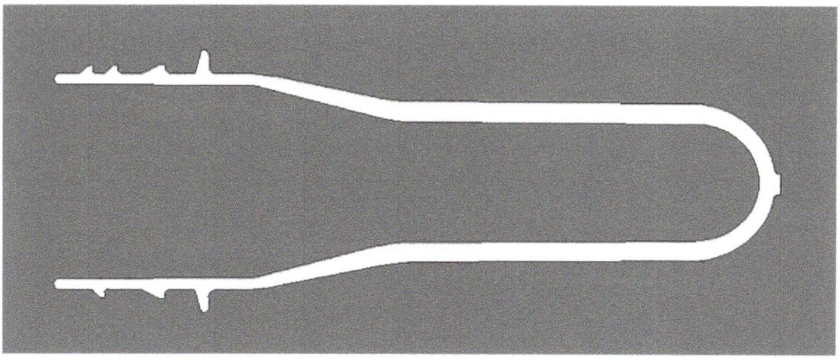

Figure 2.8 Empty cavity.

Injection molds consists of the male core, the female cavity, and the neck inserts. The latter have to move during ejection of the part to release the undercuts created by the thread beads. For this purpose they are mounted on slides that are often cam-driven. Cores and cavities are always water cooled, neck inserts may or may not. Injection molding of preforms is different from other forms of injection molding as the preform wall is relatively thick, injection pressures are relatively low, and the injection speed is low to prevent shearing of the material (Fig. 2.8).

We begin injection with the tool closed forming an empty cavity as shown earlier.

Material enters the cavity through the gate. Despite the relatively low injection pressure the material pressure may bend the injection core to one side and cause what is known as "core shift" with the resulting preform wall thickness becoming uneven. This is especially true for thin cores (below 17 mm) but may also happen for standard ones when guide bushings are worn out for example (Figs. 2.9 and 2.10).

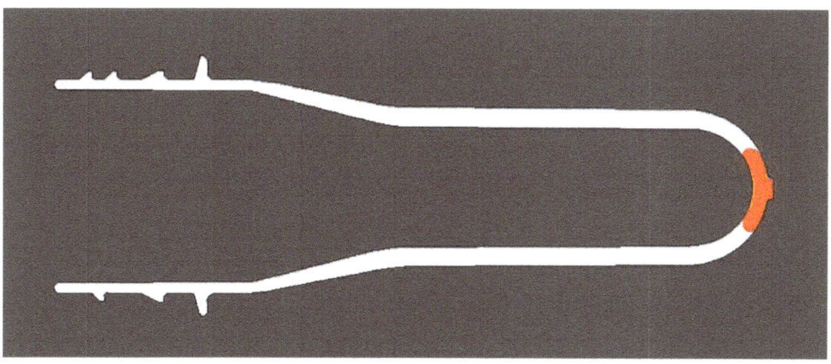

Figure 2.9 Injection 1.

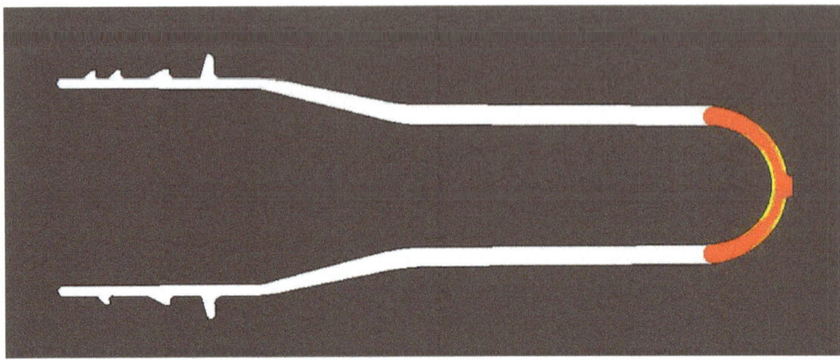

Figure 2.10 Injection 2.

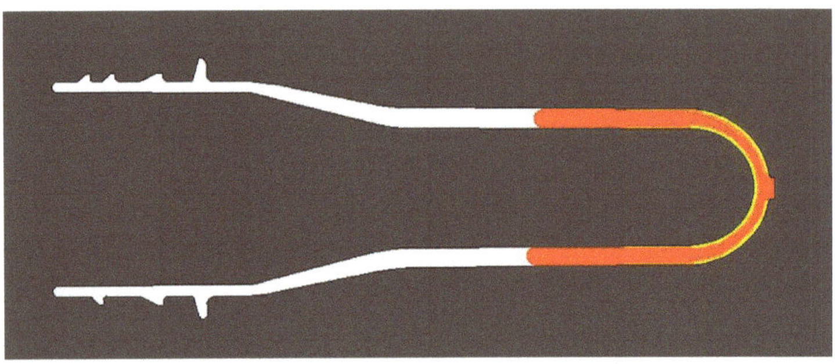

Figure 2.11 Injection 3.

As the hot material hits the cold mold walls the resin in direct contact with the wall freezes off and forms a boundary layer. The material in this layer will not change during injection. Its thickness restricts the mold channel and is one reason why a minimum wall thickness must be maintained in the preform gate area (Fig. 2.11).

As more material enters the cavity the boundary layer expands along the length of the preform. Its thickness stays the same as long as hot material is flowing through. The air that is present in the mold cavity must have an escape path. Otherwise, trapped air would lead to sink marks in the preforms. Four to eight vents approximately 0.001–0.0015-mm deep are machined into the face area of the preform neck allowing air to vent to the outside. Sink marks are also prevented and the flow of material improved by giving cores a finish in the direction of material flow rather than radially. This is achieved by special machinery that turns the cores while simultaneously moving a polishing stone back and forth on the longitudinal axis of the core (Fig. 2.12).

At this point in the injection process the cavity has been filled. The added resistance causes the hydraulic pressure to increase and it is here that the machine needs to be switched from injection to hold or packing

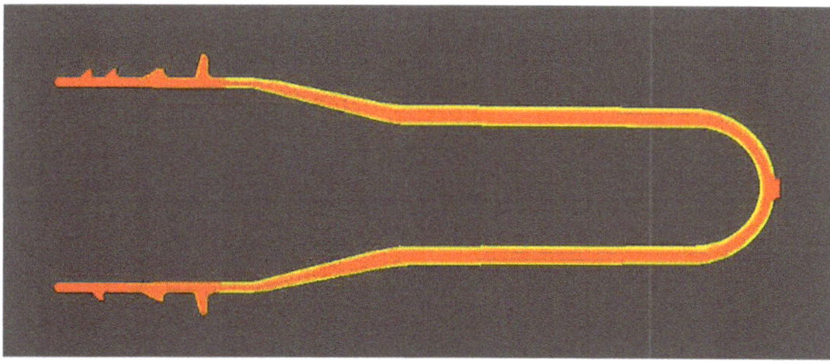

Figure 2.12 End of injection.

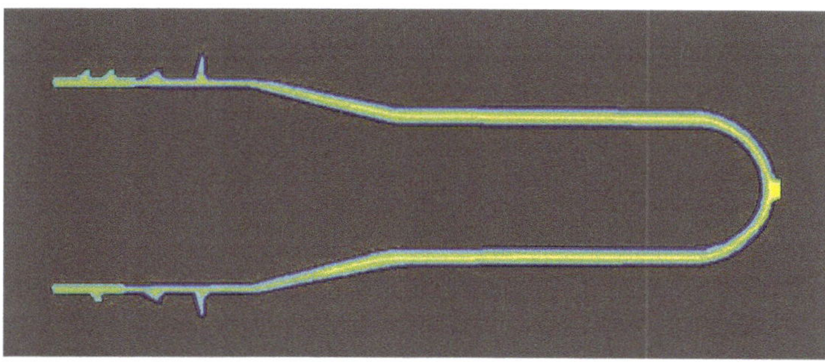

Figure 2.13 Cooling time.

pressure. This can be done by using the actual pressure as the setting to trigger the hold pressure but for PET a position-based trigger has proven to be more consistent and is therefore used almost exclusively. The point at which this occurs is called the transition or switch-over point and can be dialed in on the screen. During the hold phase material that is now starting to shrink as it cools is replaced through the still open center of the melt stream. This is necessary to avoid sink marks (Fig. 2.13).

During cooling time material is now cooling quickly and shrinking onto the core in the process. It is noteworthy that the gate area of the preform always stays warmest as it is the last part of the preform to receive hot material. Most preform defects such as cloudiness are located here for that reason. In single-stage stretch blow molding, the warmer gate area limits the processability of the preform as the temperature cannot completely be dialed in but is a result of wall thickness and injection parameters.

When problems with a particular preform arise, designers should be aware of the various aspects of the injection molding process and drying parameters and first ensure that preforms were processed correctly before making changes to the shape of the preform.

2.6 Behavior in the Blow Mold

Natural Stretch Ratio (or Natural Draw Ratio)

The stretch or draw ratio of a polymer is the ratio of the resulting length (in the direction of applied stress) to the original length. When PET is stretched, for example, during blow molding, it reaches a point at which an increase in the force is required to continue stretching. The point at which the PET requires this extra force is called the NSR for a particular set of stretching conditions.

The NSR is reached when strain (or work) hardening occurs on the stress–strain curve for materials. Recall that before a material yields, once the applied force is removed, it can return to its original dimensions. Stretching beyond the yield point results in permanent deformation, and further stretching will result in fracture. In some materials, including PET, strain hardening can occur before fracture, which is essentially the aligning (or orienting) of the structural regions of the material in the direction of the applied stress which can result in improved physical properties for the material.

The design of a PET preform is such that during stretch blow, the optimum orientation is achieved just as the stretched walls meet the mold. This point occurs just beyond the NSR. Proper stretching results in longer shelf life and less gas permeability, for example, higher carbon dioxide retention for soda. Overstretching results in a "pearlescent" appearance to the bottle signifying microcracks (fracture) and excessive deformation.

A resin with a low IV has a higher NSR than a high IV resin. The polymer chains in a low IV resin are shorter, therefore, less entangled and can be easily stretched more than a high IV resin. In the high IV resin, chain entanglement limits the amount of stretch; similar to trying to pull one end from a tangled ball of string where the knots limit the length that can be pulled out. For this reason, preform designs differ when considering low IV or high IV PET.

The following diagrams illustrate the material stretching in the blow mold without relating to actual data. Strain (elongation) is plotted on the horizontal axis and the corresponding stress on the vertical axis. To obtain these data, a heated test strip of PET might be pulled on a special machine that records the pulling force and the elongation of the strip. In the reheat stretch blow molding (RSBM) process the stretch rod and blow air provide the stress needed to transform the preform into a bottle. The top right of Figs. 2.14–2.16 indicate the stage of the preform in the blow mold.

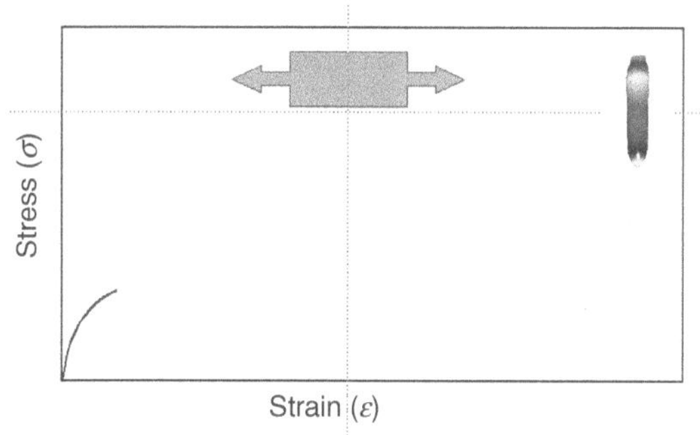

Figure 2.14 Elastic deformation occurs when the stretch rod starts moving material toward the blow mold.

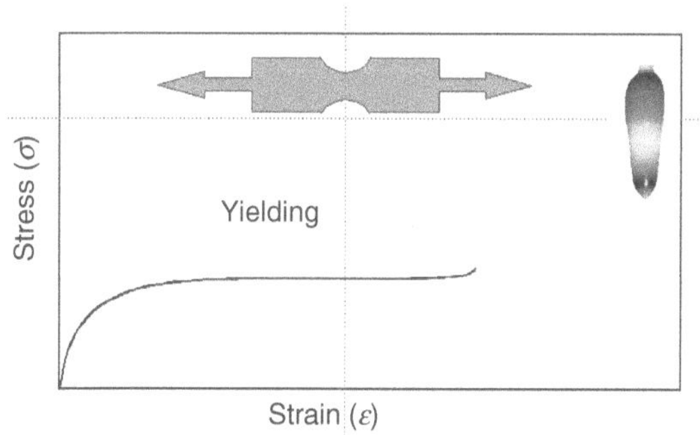

Figure 2.15 At the yielding plateau no further stress is required for additional strain.

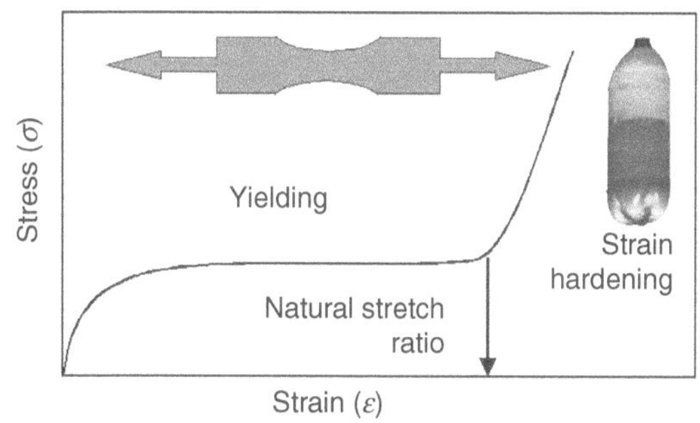

Figure 2.16 High blow pressure forces the material to strain harden.

Elastic Deformation

The first stage is the area of elastic deformation. Here the material stretches but will retract if the stress is removed. This is similar to the way that metals behave, but the shape of the curve is slightly different. Looking at a preform this stage can be compared to the stretch rod starting to push on the preform. If the stretch rod was retracted, the preform would shrink back almost to its original length.

Yielding

The second stage is yielding. With no increase in stress, the material "gives," elongating easily. This is what happens in the blow mold when the primary or preblow air partially inflates the preform. The preform will continue to inflate until it reaches the NSR, the point after which higher stress is needed to achieve further elongation.

The third stage is called strain hardening. Applied stress levels have to increase exponentially in order to force material to stretch further. This is the point in the blow process when high-pressure air enters the preform and forces it to stretch from a bubble to the blow cavity walls where it is rapidly cooled down. It is during the strain hardening phase that the material achieves orientation.

Relevant Parameters

IV, temperature, and copolymer content, all play a role in determining how far the material stretches during yielding and what force is required to stretch it further. Temperature conditioning allows the operator to improve the blow-molding process by making certain parts of the preform hotter or colder, and changing the way in which they will stretch (Fig. 2.17).

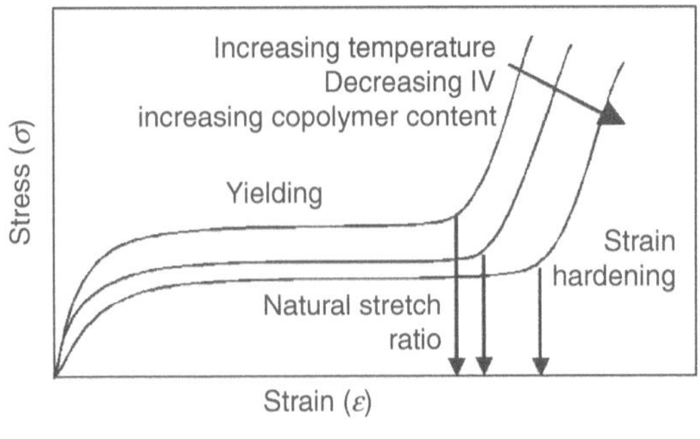

Figure 2.17 Several factors are at work determining the NSR.

Table 2.1 Virtually All Properties of PET Benefit From a High Degree of Orientation

Property	Nonoriented	Oriented
Thickness (mm)	0.36	0.36
Water vapor transmission rate ($g/m^3 \times 24$ h)	3.4	2.3
Oxygen permeability ($cm^3 \times mm/m^2 \times 24 h \times atm$)	2.9	2.2
Carbon dioxide permeability ($cm^3 \times mm/m^2 \times 24 h \times atm$)	15.7	14
Tensile modulus of elasticity (MPa)	3170	4960
Tensile stress at yield (MPa)	82	172

The objective of preform design (or selection) and blow-molding processes is to properly match up the NSR of the preform at the blow-molding conditions with the design stretch ratios of the preform/bottle combination.

Property Data for PET

Since the strain-hardening phase of the process is so important for bottle performance, correct preform design, temperature profile, and blow air timing are all necessary to guarantee the best bottle. If the inflated preform reaches the bottle mold during the preblow phase, orientation does not occur to a sufficient degree and the finished bottle might fail any number of tests (Table 2.1).

Acetaldehyde (AA) in PET Bottles

AA is a natural sweetener that is present in all citrus fruits and is often used as sweetener in beverages. It is also a by-product of heating PET, especially heating it to melt temperature. Its significance in the PET industry relates to the production of water bottles. Still water taste is very sensitive to even small AA concentrations while the sugar content of carbonated drinks and juices masks any flavor contributed by AA completely. Producers of preforms for other beverages may still monitor AA content as a way of keeping track of the maintenance state of screw and barrel because when these parts wear shear stress and with it AA creation is increased.

AA Creation

In the initial stages of PET resin manufacture, AA level may be as high as 150 ppm. What happens is that —OH end groups combine with water, glycol, or oxygen that may come in contact with the resin at that stage to

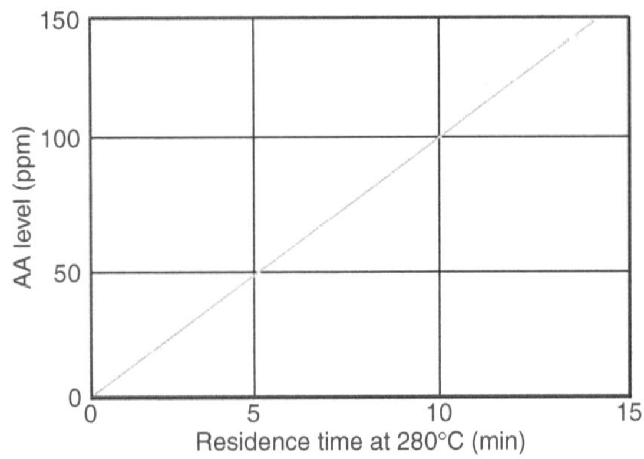

Figure 2.18 Residence versus AA level. A linear relationship exists. Diagram courtesy of Shell.

form AA whose formula is CH_3CHO. During solid-state polycondensation the material is heated and AA is removed by nitrogen gas to a level of <1 ppm.

Drying is again critical to AA generation during preform injection molding as moisture present in the resin will not only break the molecular chains but also create AA. A material temperature of 165°C (329°F) as measured at the extruder throat of the injection machine has been proven as optimal to minimize AA generation. Residence time and temperature are the other crucial factors (Fig. 2.18).

The relationship between residence time and AA level is linear. What it means for the preform molder is that extruder size and preform weight should be closely matched. One intriguing detail of PET preform manufacture is that cycle times do not vary significantly depending on weight but rather on wall thickness. Two preforms with the same wall thickness where preform A is twice the weight of preform B may differ in cycle time by only 1 or 2 s. That means the residence time of the lighter preform B is almost twice as long as that of preform A with corresponding effect on AA content. Over the past few years water bottles especially have become lighter with the negative effect that their residence time has increased if they are produced in the same extruders as they were before light weighing.

The relationship between AA generation and temperature is exponential (Fig. 2.19).

Temperature therefore plays a more important role in AA generation. To keep temperature at a minimum, processors should follow these procedures:

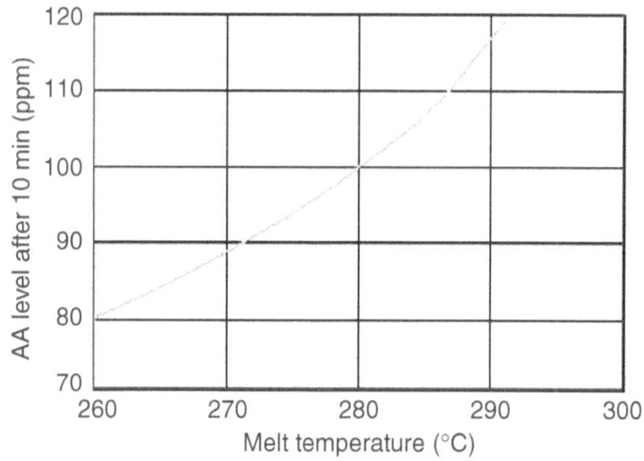

Figure 2.19 AA level versus temperature. This relationship is exponential. Diagram courtesy of Shell.

- Maximum heater adjustment should be 290°C (554°F)
- If there are percentage controlled heaters such as common for injection nozzles they should be set at a minimum and only raised for start-up
- All thermocouples need to be firmly inserted into the respective machine parts to avoid false (too low) readings
- Injection time should correspond to 10–12 g/s per cavity, that is, a 15 g preform should inject in 1.5–1.8 s to avoid shear stress
- Back pressure should be set to a minimum, again to avoid shear stress. Typical range is 10–20 bar (150–300 psi)
- Screw speed should be set to a minimum. This can be done by measuring the time that the screw is not recovering (turning) during automatic cycle. This time should be between 1 and 2 s. If it is longer, screw r.p.m. can be reduced.

Copolymers have a lower melt temperature than homopolymers (Section 2.1) and should be exclusively used for the production of preforms for water bottles. Another resin feature should be low IV because high IV increases the melt viscosity of the resin requiring higher injection force and thereby increasing shear stress. Typical IV values for water bottle preforms are 0.72–0.76.

AA in Water Bottles

The process by which AA may be released from the bottle walls into the beverage is quite complex and both humidity and temperature play a role overtime besides the initial AA level in the bottle (Fig. 2.20).

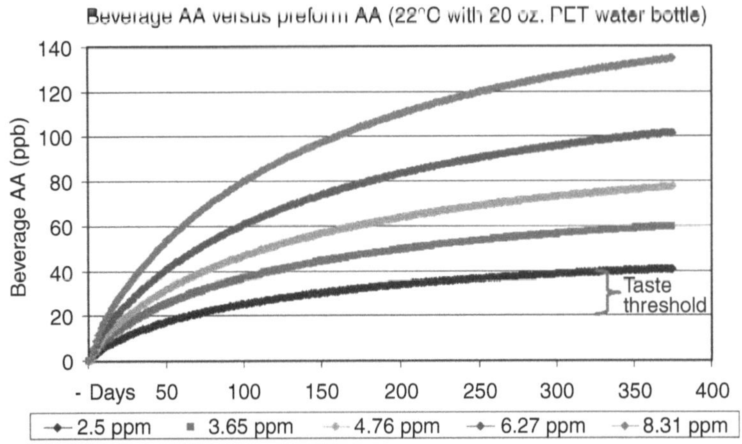

Figure 2.20 AA migration overtime.

Taste tests have shown that consumers can detect a level of 20–40 ppb of AA in their drinks. Starting with the lowest possible AA level is paramount to delay this point in time as far out as possible.

Temperature also plays a role (Fig. 2.21).

This effect can easily be experienced by leaving a water bottle on the dashboard of a car for a few weeks during the summer months. The water in this bottle will taste stale and have a light, unpleasant sweetness to it. Water bottlers have decreased the amount of AA they accept in their bottles.

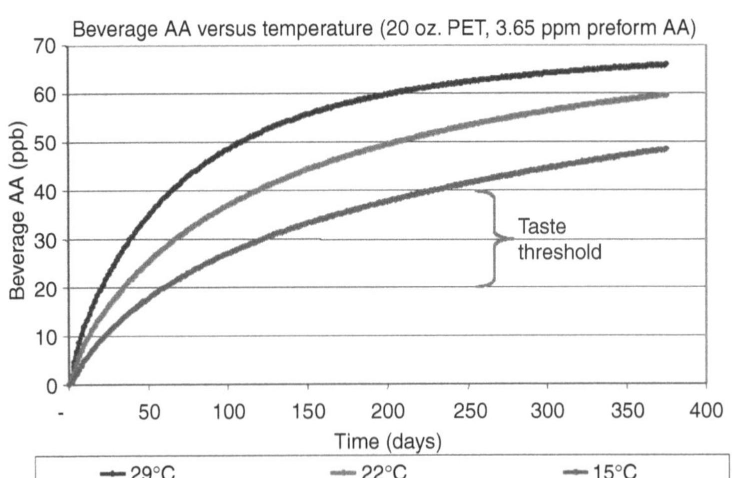

Figure 2.21 AA migration versus storage temperature.

While 10 years ago 6 ppm was widely accepted, some companies demand now levels as low as 1.5 ppm. Machine manufacturers have responded by fine-tuning screw and hot runner design and are able to deliver preforms at hose levels.

There are two ways to measure AA level. Most common is the ground-parison method. Preforms, bottles, or resin are cut into small pieces and ground up to particles smaller than 1 mm. In order to avoid creating AA during grinding specimen are cooled down with liquid nitrogen before entering the grinder. The ground material is then placed in a glass-closed vial and heated to 150°C (302°F) for 30 min. Thus the prepared sample now contains head space with AA in it and it is this gas that is being measured in a gas chromatograph.

The other method uses blown bottles. These are purged with nitrogen and stored for 24 h. Then a fixed gas volume is extracted with a syringe and measured in the gas chromatograph. It is apparent that both procedures are cumbersome and time-consuming. With the development of 144, 192, and even 216-cavity preform injection tools, it is outright daunting in its scope. A more efficient method is to check only the so-called "hot cavities." In Chapter 9, Section 9.7 I have explained the mechanics of viscous heating, the fact that the resin heats up unevenly as it moves through the barrel and hot runner channels. While this has undesired effects in injection stretch blow molding with respect to the bottles, in injection molding the effect is that the preforms in some cavities are always warmer than the majority of the others. This higher temperature also increases the AA level in the affected cavities and fortunately it is always the same cavities that show this behavior. After establishing the performance of all cavities, lab personnel can concentrate on the "hot cavities" and still make valid assumptions about the AA level of the entire tool.

3 Reheat Stretch Blow Machine (RSBM) Types

Chapter Outline

3.1 Overview

In polyethylene terephthalate (PET) bottle manufacturing reheat stretch blow machines (RSBM) are the second part of the so-called two-stage process (the first part being the injection molding of preforms). All RHSB machines use injection-molded preforms, heat them up, and stretch and blow them in a blow mold. But besides these rudimentary basics there is now a variety of machine types available for all output and quality requirements. A potential buyer should be aware of all the various differences in order to make an informed decision on what machine to purchase.

In the single-stage process both preforms and bottles are manufactured in the same machine. The single-stage process is quite different from the two-stage process in many respects. There are also machines referred to as integrated two-stage machines, which fall between the two categories. These machines injection mold preforms, transfer them to mandrels and rotate them in front of a short section of infrared lamps or other device before blowing. In Chapter 8, I will discuss single-stage machinery, integrated two-stage machinery, and conclude the chapter with a comparison between these processes and the two-stage process.

Stretch Blow Molding. http://dx.doi.org/10.1016/B978-0-323-46177-1.00003-2

Semiautomatic Machines

Most of these machines are made in Asia. They usually comprise of a stand-alone oven section, set next to a blow clamp. An operator places one or two preforms on mandrels that spin through the oven section. He also takes one or two preforms that have gone through the ovens off the mandrels and places them into the open blow mold. Typically, closing of the blow clamp requires two buttons to press in order to protect the operator from accidents. The mold closes and stretch blows the bottle(s). The operator subsequently takes the bottles out, adds preforms to the mandrels, and the process starts all over again. Outputs vary from 60 h^{-1} for 20 L or 5 gal water bottles in single cavity to 800 half liter bottles/h in dual cavity.

A slightly different version is one with four cavities and automatic feeding of the preforms into the oven system but manual placement of the preform into the blow mold after the mold has opened. The operator pushes the blown bottles out of the molds which then fall onto a conveyor underneath and the mold closes. Outputs of 1500 bottles/h are possible with lightweight bottles.

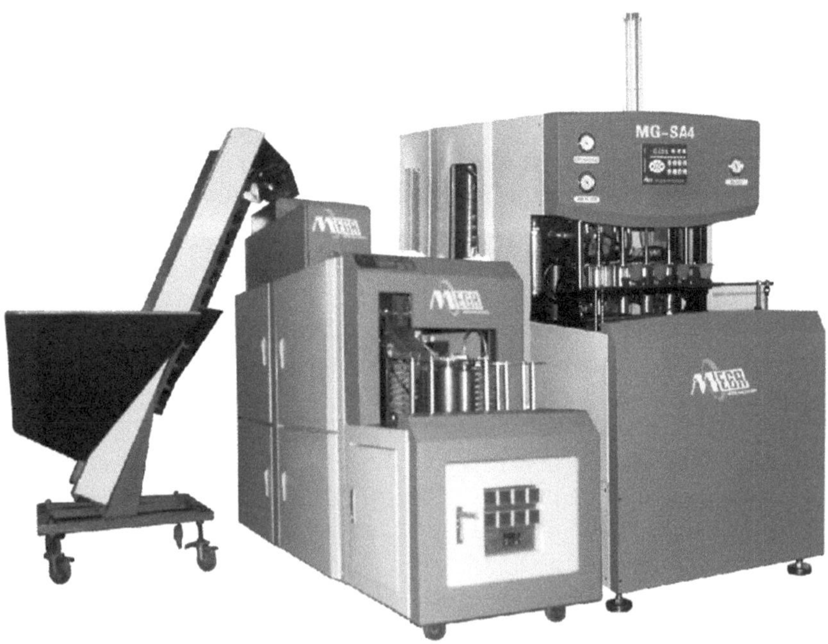

Figure 3.1 Semiautomatic machines require full-time operators but offer the lowest capital investment. Photo courtesy of Mega Machinery.

One advantage of these types of machines is that the movement of the preforms in the ovens is continuous. This reduces temperature variations between them, a distinct disadvantage of indexing blow machines (Chapter 4). One issue I have had with semiautomatic machines is control of the oven speed. Especially with thick preforms it is important to be able to control oven speed precisely as there is only a small process window between bottles breaking (temperature too low)and preforms crystallizing (temperature too high). Rotating dials are not accurate enough and should be avoided (Fig. 3.1).

Most machines are sold with at least one mold, a compressor, and possibly a small chiller. Quality varies between manufacturers and buyers should vet each supplier or go through a distributor to avoid disappointment. More will be discussed in Chapter 11 (Fig. 3.2).

Figure 3.2 Five gallon bottle. It is due to the low volumes 5 gal bottles are often manufactured on semiautomatic machines. A handle is placed in the mold before blowing. Photo courtesy of Mega Machinery.

Linear Shuttle Type Machines

In these machines blow molds are mounted together and all move with a common cylinder. Cavitations of up to eight cavities have been built but the most common models have one to four cavities. Thus these machines cover the lower output range of applications, often competing with the single-stage process (Fig. 3.3).

Preforms travel from a common hopper to an unscrambler via an incline conveyor and fall between two rotating and inclined rollers so that gravity forces them to slide down the rollers supported by their transfer ring. A rotating flap prevents unscrambled preforms from reaching one (or more) rails where preforms now hang and glide down toward the pick-up station. Preforms are still right side up and are usually turned upside down via a

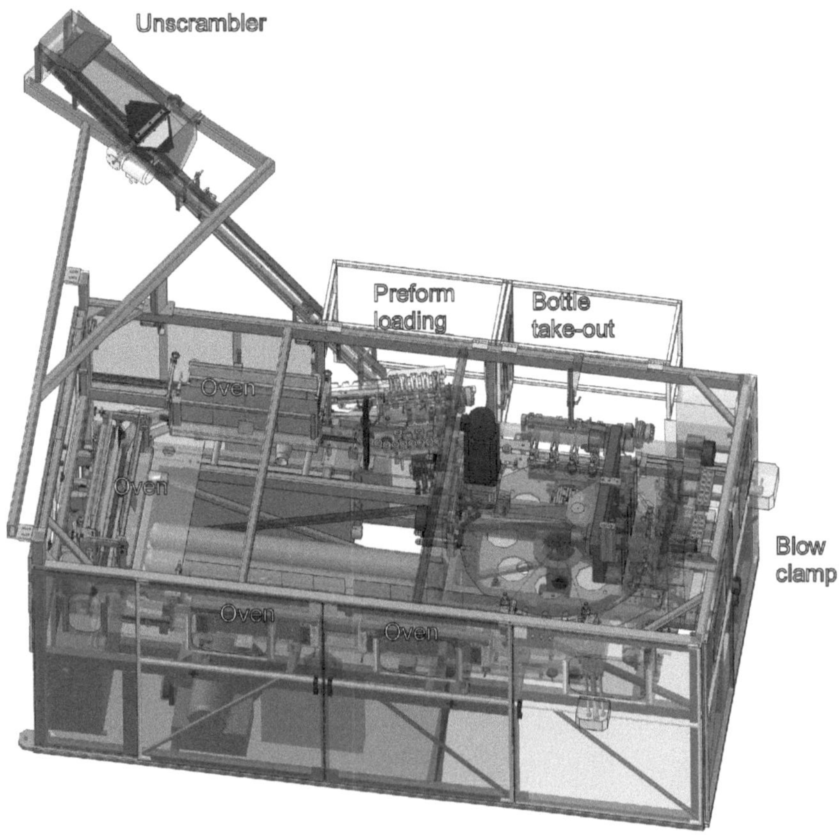

Figure 3.3 Layout of linear, five-cavity machine. Diagram courtesy of Amsler Equipment Inc.

Figure 3.4 Most preforms are blown upside down on this type of blow molding machine. Photo courtesy of Chumpower Machinery Corp.

pick-up device that consists of the appropriate number of grippers driven by three, pneumatically driven cylinders.

The linear machines then index and grippers deliver the appropriate number of preforms in one motion to the waiting mandrels. Most mandrels are made of aluminum with a diameter just below the minimum inside diameter of the preforms.

Referring to Fig. 3.3 preforms follow the mandrel track counterclockwise around the machine from the feeding station through the oven section. Equilibration happens in the turnaround section before entering the five-cavity blow clamp. Five mandrels are shuttled at the same time into the blow mold where the preforms are stretched and blown at the same time (Figs. 3.4 and 3.5).

All blow cavities are side by side. On some machines both mold halves move with a common cylinder, other machines feature individual cylinders. In most machines preforms travel upside down and the stretch rods are engaged from the bottom. After blowing, the five bottles are shuttled to the take-out station where they are turned upright onto a conveyor belt. Stand-up conveyors are common for the range of outputs possible with these machines (Fig. 3.6).

While these machines also offer a low entry level into the PET market, they do have some serious drawbacks. The most serious is the way preforms go through the oven system. It is nearly impossible to have all preforms spend exactly the same amount of time inside the ovens as they are being indexed. This becomes more problematic as cavitation increases. Furthermore, the heat an infrared lamp emits is not even along its length (Fig. 3.7).

The center of the lamp emits usually 10–20% more heat through higher temperature than the ends. The result of both circumstances is

Figure 3.5 One electric servo drives all molds on this machine. Photo courtesy of W. Amsler Equipment Inc.

that preforms are not evenly heated up and temperature differences as high as 8°C (14°F) between preforms in a four-cavity machine are not unusual. This in turn leads to differences in bottle wall distribution. Some of these differences can be improved on if each cavity has a separate set of blow valves but not all machines are equipped with this feature. Some machines feature oven tracks where preforms run parallel to each other with lamps in between. These work better but are still not at the level of continuous motion machines that will be discussed later. Most machines also space the preforms in the same pitch as the blow cavities. This leads

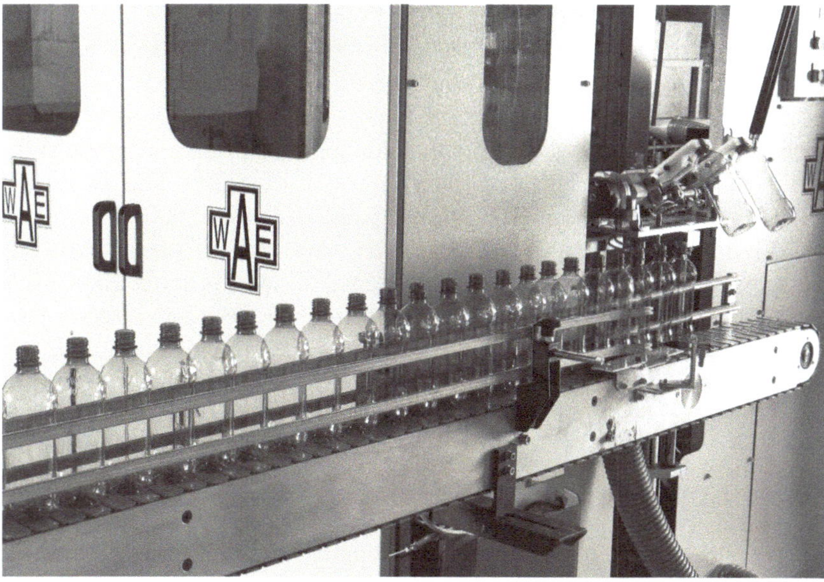

Figure 3.6 Stand-up conveyors are most commonly used at the end of linear machines. Photo courtesy of W. Amsler Equipment Inc.

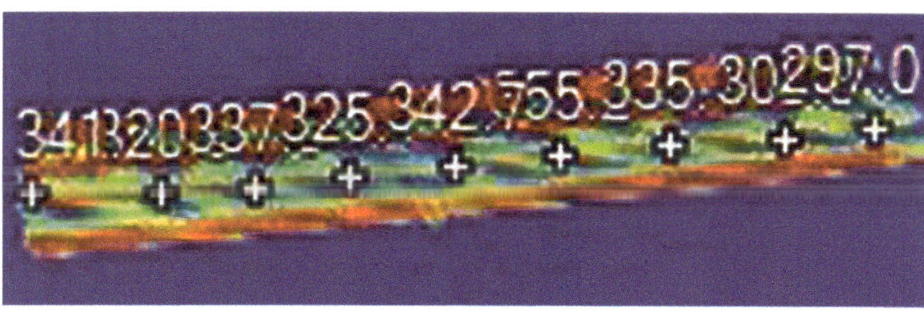

Figure 3.7 Infrared lamps do not output heat consistently over their lengths as seen in this infrared photo.

to an uneconomical use of oven power as the lamps heat up a large amount of air. This heated air in turn creates a noise factor with respect to consistent preform heating. As most blow molding plants are not air-conditioned, temperature variations have a greater impact on the heating profile of the preforms when there is so much air inside the oven system relative to the number of preforms. Many machines offer a close looping of total oven output to measured preform temperature but they do not catch all the changes that happen in a typical day/night temperature cycle. Another downside of these machines is the rotating movement of first the preforms (upside down) and then the bottles (right side up). The mounted devices with three cylinders each (up/down, forward/backward, turn in/turn out) can be a maintenance issue that has plagued many companies.

The simple design of the mandrels makes them easy to repair or replace but does not give the same security in holding the preforms concentric as the more sophisticated designs of different types of machines. This is because the mandrel outside diameter has to be about 0.02 mm (0.0001 in.) smaller than the lower tolerance dimension of the neck inside diameter but many preforms will be larger as the tolerance range allowed is about ±0.008 mm (0.003 in.) for the most common neck finishes. In the worst case scenario preform necks are 0.18 mm (0.007 in.) larger than the mandrels and, as a result, preforms wobble and are heated unevenly on their way through the ovens. This can lead to wall thickness variations from one side to the other.

Linear Continuous Motion Machines

To improve the performance and quality of low-to-medium output machines companies have developed a new type of machine borrowing some proven concepts from the rotary line of blow molding machines. This starts with the in-feed section. On these machines only a single rail is

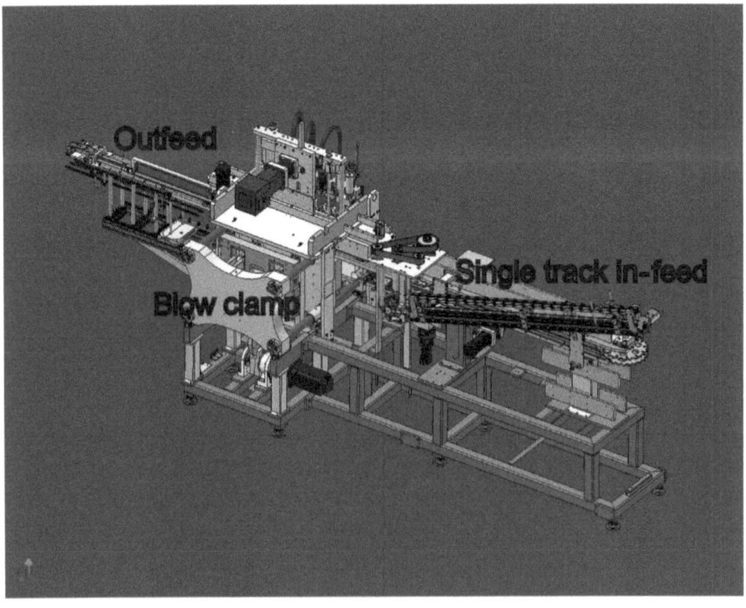

Figure 3.8 Linear continuous motion machine. These machines have gained popularity as they combine excellent quality with a lower price tag compared to rotary machines. Diagram courtesy of SIAPI.

Figure 3.9 The in-feed section of these machines is very similar to that of rotary machines. Picture courtesy of SIPA.

mounted and preforms flow continuously into a star wheel. They are right side up at this point and will stay this way throughout their journey through the machine (Figs. 3.8 and 3.9).

As the star wheel turns, mandrels move down into the neck opening of the preforms and pick them up via a spring-loaded mechanism. Spacing of the mandrels is around 38 or 50 mm, allowing necks with maximum transfer rings of 36.5 and 48.5 mm diameter, respectively. There are a few machines for preforms with wider necks that will be described later. Close preform spacing improves oven efficiency and the continuous motion guarantees even preform heat. Once preforms leave the oven section, they are grouped in the appropriate number of cavities (up to 16 at the time of this writing) and shuttled into the blow molds where they are stretched and blown together (Fig. 3.10) While they travel into the mold area, the pitch is increased to the cavity pitch by a pitch-changing device, which is proprietary to each manufacturer. Each cavity has its own set of blow valves, allowing fine-tuning and reducing dead-air volume. The same device that

Figure 3.10 Blow clamp and mold. This machine uses a cam-driven blow clamp and base insert mechanism controlled by a servo motor. Picture courtesy of Chumpower Machinery Corp.

shuttles preforms in also takes the blown bottles out either on stand-up or air conveyors.

Machines of this type use servo motors to achieve proper synchronization between the mandrels that move continuously and the pick-up devices. Low energy consumption and little maintenance are resulting benefits. While the time between the end of oven section and the start of blowing is slightly different for each preform (the first preform may exit the oven section up to 3 s before the last) I have not found this to be of any consequence to the bottle wall distribution and never had to use individual blow air control. This control is still valuable because it allows placing the blow valves close to the cavity thereby reducing the "dead air" volume, that is, the volume of air that does not contribute to the blowing of the container but must be supplied and exhausted with every machine cycle.

The great advantage to these machines is that they deliver nearly identical quality as rotary machines at outputs of up to 1800 bottles/cavity per h but at a much reduced capital cost. Rotary machines are more expensive because of the costly distributor systems that are in the center of the blow wheel. The excellent quality of the linear continuous motion machines stems from the fact that all preforms receive the same heat and can be

Figure 3.11 Some machines pick preforms right side up then turn them for their travel through the oven system. Picture courtesy of UAB Terekas.

blown into bottles with identical wall thickness distribution. One caveat for owners of rotary machines may be that molds of those may not fit into some linear machines (Fig. 3.11).

Rotary Machines

These machines are the stars of the blow molding world and produce most the world's soft drinks and water bottles. This market segment consumes up to 80% of all PET converted into bottles and the resulting income stream has enabled manufacturers to pour significant amounts of money into R&D efforts. This in turn has resulted in massive innovation driving outputs up. Output is usually measured in bottles/cavity per h or b/c per h and this number has increased from 800 in the 1990s till today on some machines. Machines up to 40 cavities and with outputs of up to 80,000 bottles/h are on the market today. The enormous progress was possible by increasing mechanical speeds and fine tuning process time. One major obstacle to running at the highest output rates is the thicker wall thickness in the base of a bottle that needs cooling time. Readers should note that published output numbers always refer to lightweight water bottles that feature completely stretched-out bases. Custom and carbonated soft drinks (CSD) bottles require longer process time and thus reduce machine output.

The oven sections of rotary machines are very similar to those of other machine types and each manufacturer has a particular design to effectively heat preforms and avoid undue temperature increase in the preform necks and at the connections of the infrared lamps. More on that is in the section on oven design (see Chapter 4).

The main difference to other machine types is in the design of the blow clamp. In a rotary machine each preform travels to its own blow clamp that is arranged on a blow wheel. All services to these clamps come through the center of the blow wheel by means of rotary distributors. A massive drive gear sits underneath the wheel and a variable speed motor drives it. The time between end of oven section and start of blowing is identical for each preform and this feature allows rotary machines to deliver excellent quality very consistently. The single electric motor drives all machine functions of the machine via timing belts. Mechanical cams operate stretch rods and mold-opening and -closing sequences. Therefore, to change speed it is only necessary to change the motor speed and all other functions are automatically synchronized (Fig. 3.12).

The layout given in Fig. 3.13 will familiarize the reader with the main machine components.

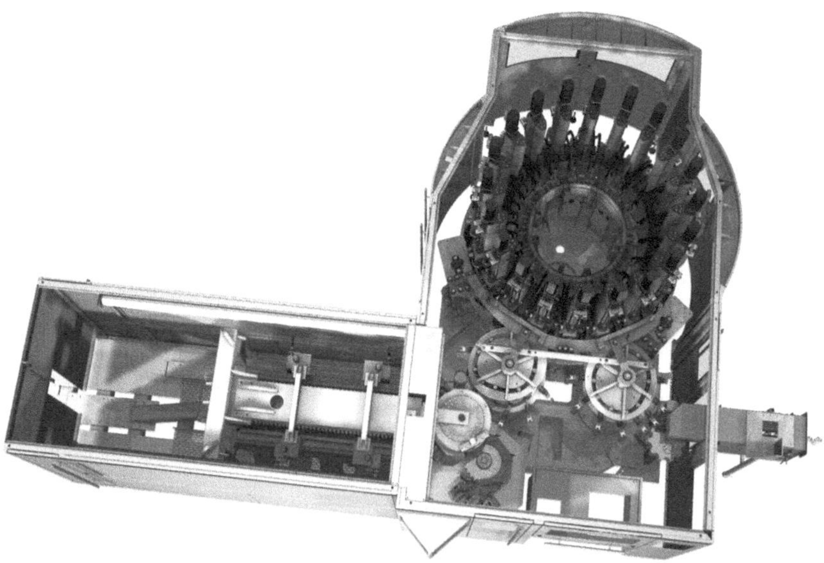

Figure 3.12 Blow machine top view. Linear ovens and blowing wheels have become standard in the design of rotary machines. Photo courtesy of KHS Corpoplast GmbH.

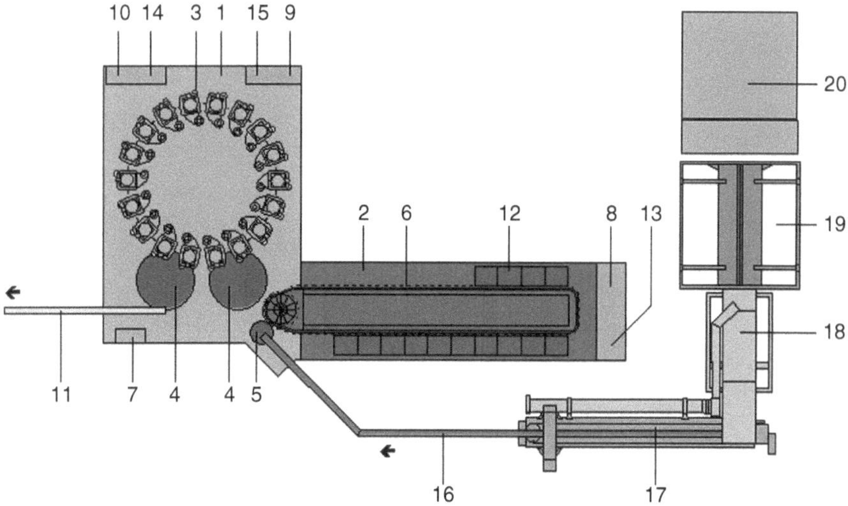

Figure 3.13 Layout of typical rotary blow molding machine. *1*, Blowing module; *2*, heating module; *3*, blowing unit; *4*, blowing unit for in-feed and discharge star wheel; *5*, in-feed star wheel; *6*, mandrel chain; *7*, control panel; *8*, control cabinet; *9*, water; *10*, pneumatics; *11*, air conveyor; *12*, heating unit; *13*, electricity supply; *14*, air supply; *15*, water supply forward flow/backward flow; *16*, preform in-feed rail; *17*, preform unscramble; *18*, preform incline conveyor; *19*, preform storage bin; *20*, preform unloader. Diagram courtesy of Krones AG.

Preforms enter the machine at position #20. Injection molders often pack their preforms in so-called "gaylords" cardboard storage boxes of roughly 1 m^3 volume. Resin companies use these gaylords to ship resin and they are widely available. The unloading station might have a "gaylord tipper" that enables personnel to easily tilt gaylords to empty them. Position #19 is the storage bin, a simple metal enclosure open to the top (Fig. 3.14).

Position #18 is the incline conveyor that takes the preforms to the unscrambler. Incline conveyors are controlled by downstream switches that detect the amount of preforms in the line and only feed preforms when needed. There are typically two switches. The bottom one turns the machine in-feed off and triggers the alarm when no preforms are detected. The top switch turns the incline conveyor off.

Position #17 is the unscrambler. It orients the preforms onto the in-feed rails position #16 with flexible paddles and the support of gravity. The preforms hang from their neck support ring between two stainless steel rails and travel this way again by gravity to the in-feed star wheel at position #5. Some preforms will fall off the rollers and these are fed back to the storage

Figure 3.14 Storage bin and incline conveyor to unscrambler. Photo courtesy of Krones AG.

bin. To avoid unnecessary scratching the incline conveyor speed that has a variable-speed drive should be low enough so that few preforms have to go back to the bin (Fig. 3.15).

In the in-feed section preforms are transferred by various means onto mandrels (Fig. 3.16).

These mandrels feature various spring-loaded devices to hold the preform in place and guarantee both alignment and secure assembly. Most mandrels ride in tracks (position #6) and feature a sprocket on the top (in oven systems where the preforms ride upside down, the sprocket is at the underside). The sprocket engages with a chain that spins the mandrels and may also move the mandrels through the oven section position #2. Spacing between mandrels on the track ranges from 38 to about 50 mm (1.5–2 in.) and can be increased to double that figure for wide-mouth applications. The majority of soft drinks and water bottles use necks of 28–33 mm diameter, easily accommodated by the standard spacing (Fig. 3.17).

Spinning with the mandrels, the preforms are exposed to infrared radiation. The machine shown here has 14 heating ovens (position #12), 10 on the way away from the blowing wheel position #1 and 4 on the return route. In the turnaround section as well as in between the last oven in the line and the blowing wheel preforms undergo equilibration, that is, balancing of temperature differences inside the preform wall (Chapter 6) (Fig. 3.18).

Figure 3.15 Linear oven with rail. Preforms hang from their neck support rings on rails (upper left of the picture) on their way to the in-feed of the machine. Photo courtesy of Krones AG.

Figure 3.16 In-feed section. This machine uses a star wheel to transfer preforms onto mandrels. Photo courtesy of KHS Corpoplast.

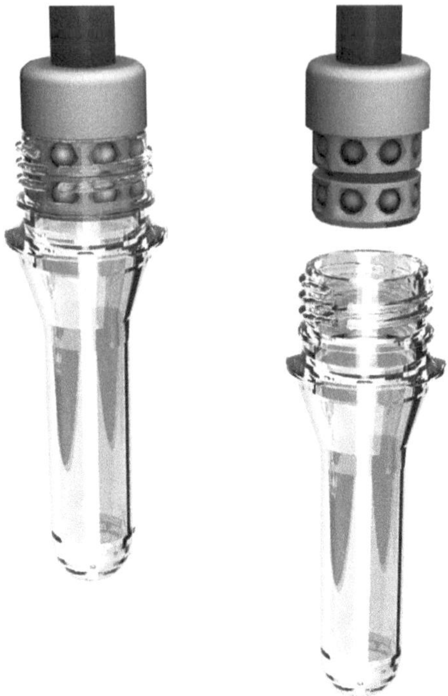

Figure 3.17 Mandrel uses an innovative design with two rows of balls holding preforms firmly in place. Photo courtesy of KHS Corpoplast.

Figure 3.18 Preforms in oven. In this machine preforms pass through the ovens upside down. Photo Courtesy of Sidel Inc.

At the end of the oven track preforms enter a star wheel or other transfer device and are moved one by one onto the blowing wheel position #1. The machine described has 18 blow cavities position #3. Each cavity has its own blow mold, stretch rod, water connection, and three blowing valves for preblow, blow, and exhaust. Air and water connections as well as electrical lines for sensors and switches come from a feeding unit located in the center of the blowing wheel. These central distributors are one of the reasons that rotary machines are more expensive compared to linear machines of the same cavitation (Fig. 3.19).

Figure 3.19 Blow mold half and bottle shown here is one of the two blow mold halves with bottle and bottom insert. Photo courtesy of KHS Corpoplast.

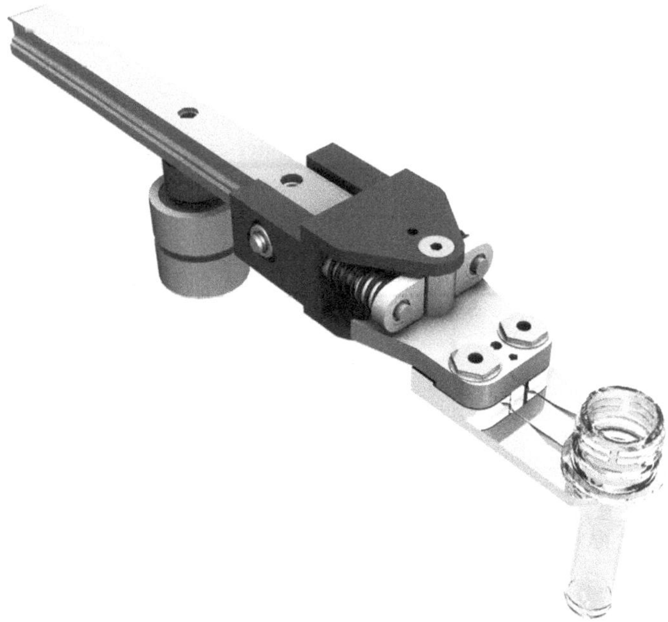

Figure 3.20 Gripper. One of many gripper designs. Diagram courtesy of KHS Corpoplast.

Blown bottles leave the mold via grippers or star wheels onto conveyors, position #11. These grippers are very different from the ones used on linear machines and transfer preforms between ovens and blow molds and bottles between blow molds and discharge unit. They are cam-controlled and rotate on shaped tracks. Instead of having cylinders that open and close them, they are actually just pushed and pulled into and from preforms that are securely held in star wheel pockets or the blow molds in the case of finished bottles. This saves time and maintenance. It does require a particular preform neck design (Fig. 3.20).

To facilitate the handshake between star wheel and gripper there must be room for the gripper to engage the preform above the transfer ring as the preform is hanging from it in the star wheel. This requires the preform to have a straight section underneath the thread beads for the gripper and another section underneath the transfer ring for the star wheel (Fig. 3.21).

Brand owners in industries such as cosmetics may find transfer rings unappealing for their customers and it is possible to run preforms without a transfer ring through a RSBM machine albeit at a slower speed.

High-speed machines generally use air conveyors, where bottles hang from their neck support ring, gently pushed by air blowers. Antislip agents in the resin help reduce any friction in the conveyor lines.

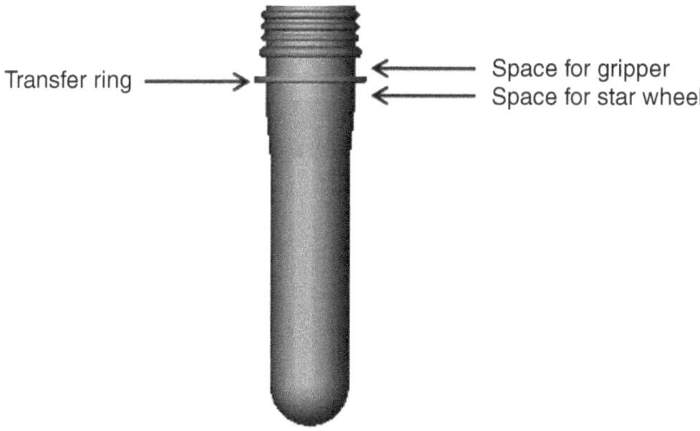

Figure 3.21 Preform during handshakes between various parts of the machine the area above and underneath the transfer ring is used for grippers.

3.2 Differences Between Rotary Machines of Different Manufacturers

Differences between rotary machines include the following:

- the orientation of the preforms in the ovens and blowing section
- the number of cam-driven movements
- the shape and location of the oven section
- the way the blow molds move
- the way stretch rods are driven
- the operator interface
- various process options.

3.3 Orientation of Preforms and Bottles

In some machines preforms travel neck down through the ovens, in others neck up. Neck down orientation was thought to make it easier to protect the neck from the heat of the infrared lamps and was at one time the preferred method. However, advances in neck protection through more elaborate air flow systems has removed this obstacle and two of the three most popular machines now use the neck up position in their machines. This has eliminated turning of preforms and bottles and has thus simplified machine layouts and led to gains in operational efficiencies. Some manufacturers of linear machines also have adopted this preform transport method, which seems to become the standard. From a process point of view there is no difference in the orientation of the preform during blowing and there is slight advantage of the neck up blowing position with pneumatically operated stretch rods as they can move faster when they work with gravity instead of against.

3.4 Movement Actuation

Most rotary machines have some movements driven by mechanical cams while using cylinders for other movements, and vary in how these two methods of actuation are used. Cam-driven movements are more difficult to adjust but tend to be very stable without variations in speed or force. They also need little maintenance for long periods of time. Cylinder-driven movements are more flexible and operators can easily adjust them from the control panel of the machine. However, they tend to show slight variations in movement behavior and require more frequent maintenance. Servo-driven stretch rods are now in use by some manufacturers offering consistent performance and ease of setup through the operator interface. Stretch rod adjustment can be done in a number of ways with different stops for each bottle. For example, a servo motor system can do this automatically and therefore reduces change overtime.

3.5 Shape and Location of Oven Section

The trend over recent years has been to design machines that have U-shaped oven tracks. There are no ovens in the turnaround section, thus giving some extra equilibration time (Chapter 6). Ovens can easily be accessed and the distance between lamps and preforms is equal over the whole length of the lamp. Round oven sections were used on some machines at one time but are no longer manufactured except with one particular model that uses microwaves instead of infrared.

3.6 Blow Mold Actuation

Most molds open horizontally and have some type of locking mechanism. They differ in the way that the cylinders and locking pins are mounted and whether or not they have so-called pancake cylinders for pressure compensation (Chapter 4). Most mold movements are cam actuated because their movement never changes no matter what bottle is running (Figs. 3.19 and 3.22).

Figure 3.22 Blow clamp in closed position with vertical locking pin engaged. Picture courtesy of Krones AG.

3.7 Preform Seal

When blow air enters the preform all blow machines must prevent this air from getting out. There are several ways of doing this with O-rings at the bottom of the preform seat or in the sidewalls of the male connection piece. Distortion of the neck may occur when the neck is getting too warm and the material loses some of its strength or when the hoop stress (increasing with the square of the diameter) becomes too big even for a cold neck.

A more elegant solution is the so-called blow dome. Instead of sealing somewhere on the inside diameter of the preform this device seals on the transfer ring. Thus by having air pressure on both the sides (ie, inside and

Figure 3.23 Clutches such as this model are used to protect fast-moving parts of blow machines. Drawing courtesy of Mayr Corporation.

outside) of the neck finish there is no net force on the neck itself. This virtually eliminates any distortion problems and leaves neck as they come from the injection machine.

3.8 Synchronization and Crash Protection

The main advantage of rotary machines is that the movement of all parts is continuous. One motor drives the entire machine and transmits movement through belts and pulleys. As long as belts are at the right tension (a maintenance must do) synchronization is assured. But what happens when a misformed preform prevents the blow mold from closing? At speeds of 2000 b/c per h serious damage would occur if there was no protection. This protection is afforded by a series of clutches that are mounted to every drive pulley that is connected to a rotating machine part. When the torsion force on the clutch exceeds a fixed value, the clutch severs the connection between drive pulley and machine part allowing free rotation. After the operator clears the jam he/she can manually rotate the machine part back into the synchronized position on the clutch (Fig. 3.23).

4 Machine Details

Chapter Outline

4.1 Oven Section

The importance of a properly designed oven section cannot be overstated. This refers to both the overall layout and the design of the actual ovens. Ovens must:

- impart the correct temperature profile into the bottle wall
- give processors flexibility in heating sections of the preform to the best temperature for even bottle wall thickness distribution
- have large enough blowers to control heat buildup inside the ovens
- protect the neck finish of preforms
- cool the infrared lamps for maximum life-span
- cool the reflectors opposite the lamps
- leave room for sufficient equilibration time
- be sized for the correct throughput at a given preform wall thickness.

Stretch Blow Molding. http://dx.doi.org/10.1016/B978-0-323-46177-1.00004-4

It is apparent from this list that oven design has attracted a great deal of engineering effort and modern machines feature excellent systems that are able to deliver lightweight bottles conforming to ever-increasing specifications.

Layout

As previously mentioned preforms may travel through the oven system neck-up or neck-down. As long as heated air is efficiently vented out of the system both methods work well.

Fig. 4.1 is a layout of a neck-up system.

There are usually eight or nine infrared lamps spaced vertically at a distance of 15–19 mm (5/8–3/4 in.). Close spacing is necessary in order that the corresponding sections of the preform can be heated exactly to the point where the best bottle wall thickness distribution is achieved. Older designs incorporated larger or twin lamps that did not offer processors enough flexibility and these systems are no longer being manufactured.

Lamps should be adjustable horizontally. This allows the processor to move them closer into areas that need the most heat. This is usually the section immediately below the neck. This area requires more heat as it is adjacent to the cooler (unheated) neck, and also because the shoulder of the bottle requires less material in most cases. Failure to heat up this section sufficiently results in a ring of thick material in the bottle that is both unsightly and leaves less material for the rest of the bottle where it is needed. Lamps are also often pushed outward to heat the gate area of the preform from above. In this case the lamp may be rotated in order that it

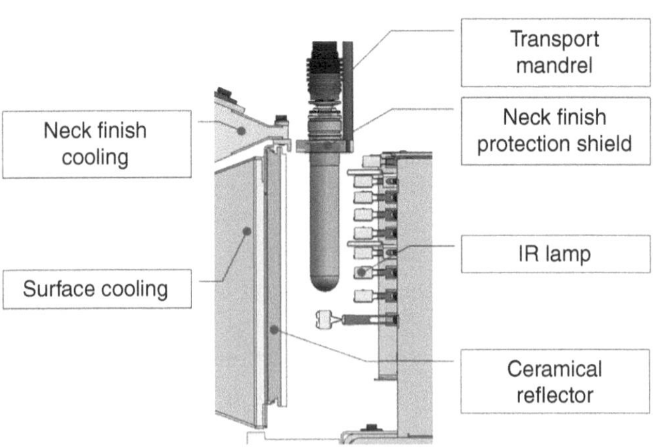

Figure 4.1 Layout of infrared oven. Diagram courtesy of Krones AG.

Figure 4.2 Easy access to the infrared lamps is another important design feature. Courtesy of Krones AG.

points downward. Another design feature is the use of tubes with a higher linear watt density in the first (or the first two) row of lamps. Wattages range from 1000 to 3500 W.

The stack of lamps must also be adjustable in the vertical direction to allow them to be aimed at the exact location in preforms that may vary with the type of neck finish used (Fig. 4.2).

Opposite the lamps are the reflectors. These are, of course, made from highly reflective material and contribute up to 40% of preform heating. Polished aluminum has proven to be the most economical material to use. To prevent oxidation, reflectors must be cooled either by air or water and both methods are in use. Tarnished reflectors must be replaced or resurfaced, otherwise energy consumption will increase and/or bottle quality will suffer.

Protecting the neck finish from infrared heat is another crucial part of oven design. Polyethylene terephtalate (PET) starts getting soft at temperatures above 65°C (150°F). If necks soften they are subject to deformation under the impact of hoop stress, that is, the force of the air pressure against the inner neck walls or through handling inside the machine. Neck finishes are protected with one or several metal shields that are either air or water-cooled.

Hoop stress increases with the diameter of the neck. Therefore widemouth bottles are exposed to higher hoop stress. Air cooling may be

Figure 4.3 An infrared temperature sensor is mounted before the blowing section monitoring preform temperature.

sufficient for a 28-mm neck but water cooling may be necessary to protect a neck of 63 mm.

At the end of the oven track, just before preforms enter the blow station or blow wheel, a single infrared thermocouple provides information on preform skin temperature (Fig. 4.3).

It should be noted that the thermocouple by its nature cannot provide information on temperatures underneath the preform skin. Depending on preform thickness and equilibration time the sensor might report the same temperature for very different preform conditions. Since it is only monitoring one specific location, the reading will only change when the particular lamp or lamps that are affecting this spot are changed. This can lead to confusion and operators should take note which lamp is affecting the part of the preform that the sensor is monitoring. However, the reading is very useful in monitoring the process once it has been established and machines allow the setting of upper and lower tolerances and will alert operators to deviations. Some machines also close-loop the preform temperature with the lamp output thus reducing the impact of changing environmental conditions. Because of complex interrelationships, algorithms employed on some older machines do not always result in the most stable process and many operators choose not to use this option.

Infrared Lamps

The lamps themselves consist of tungsten filaments sealed in quartz tubes. Quartz is transparent to infrared radiation and can withstand the high temperatures that the lamp generates. The filament temperature determines the wavelength emitted (Table 4.1).

The tubes are filled with halogen and an inert gas to avoid oxidation and blackening of the tube as well as lowering the operating temperature to about 800°C (1472°F). Lamp emission peaks at 1200 nm, which is the optimum for PET processing (Chapter 6) (Fig. 4.4).

Pinch sections of the lamp, where the tungsten filament connects to the electrical supply, must be kept below 350°C (662°F), otherwise the molybdenum film at the contacts starts to oxidize and the pinch may actually crack, causing the lamp to leak. Most ovens have blowers at the back of the lamps accomplishing this task and also monitor the temperature with special thermocouples.

Bare hands should never touch lamp surfaces, as oil can be transferred on them that leads to burn marks. Should it happen, lamps should be wiped off with alcohol.

When it comes to selecting lamps from different suppliers one important detail is how much of the applied energy is actually converted to infrared

Table 4.1 Relationship Between Operating Percentage, Used Voltages, and Temperature Output of a Typical Infrared Lamp

Operating Percentage	Respective Voltage (V)	Temperature (K)	Temperature (°C)	Temperature (°F)
40	88	1800	1527	2780
70	154	2000	1726	3140
100	220	2600	2327	4220

Figure 4.4 Typical infrared lamp. Photo courtesy of Philips.

radiation. This has a significant impact on energy consumption and buyers should carefully scrutinize data supplied by manufacturers. Half-round reflectors at the back of the lamps have proven to be a significant improvement in this respect.

It should also be noted that although these lamps emit a wide spectrum of light waves, only some of which are optimal for PET (Chapter 6). Near-infrared (NIR) heaters concentrate more of the output in the 1000-nm range, which is especially suitable to penetrate preform walls. Ovens with these heaters can reduce heating time, space requirements, and energy costs, and manufacturers have started to equip their ovens with them.

Different Heating Methods

A completely different heating system has now been developed that uses microwave beams instead of infrared lamps (Fig. 4.5). Heating time is said to be around 3 s, allowing the use of just 16 heating mandrels for an eight-cavity blow molding machine compared to about 160 mandrels in a conventional oven system. This significantly reduces changeover time for a different neck finish. Preforms absorb 95% of the microwave energy with excellent penetration independent of their color, making fan cooling and

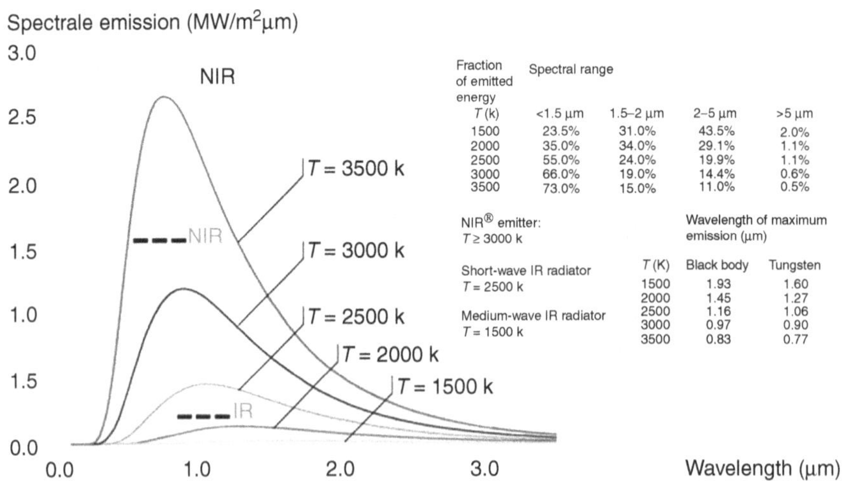

Figure 4.5 Standard infrared emits favorable wavelengths below 1.5 μm (1500 nm) only at high output rates of 3500 K whereas NIR delivers these over a larger range. Diagram courtesy of Adphos Group.

Figure 4.6 This system uses a rotary microwave oven system that drastically reduces heating time. Picture courtesy of Krones AG.

equalization time a thing of the past. Microwave stations are mounted in a rotating oven system rather than a linear one making the oven section very compact. Another advantage is that the neck finish does not need protection opening the door for improved wide-mouth container processing. A possible disadvantage is that profiling the heat output is not possible. This means that the preform design has to be perfect so that an evenly heated preform yields a bottle with even wall thickness. This is an exciting step into lower energy consumption and better control over preform temperature and we will see how far it can go (Fig. 4.6).

Another new approach is heating with laser beams. Beams created by diode lasers are emitted in the NIR spectrum directed very precisely onto the various areas of the preform with limited convectional heat. This oven system can be linear or circular as shown in Fig 4.7.

Lamp Control

There are a variety of control mechanisms present depending on machine manufacturer and year of manufacture. Modern machines allow adjustment from control screens while older machines use a number of

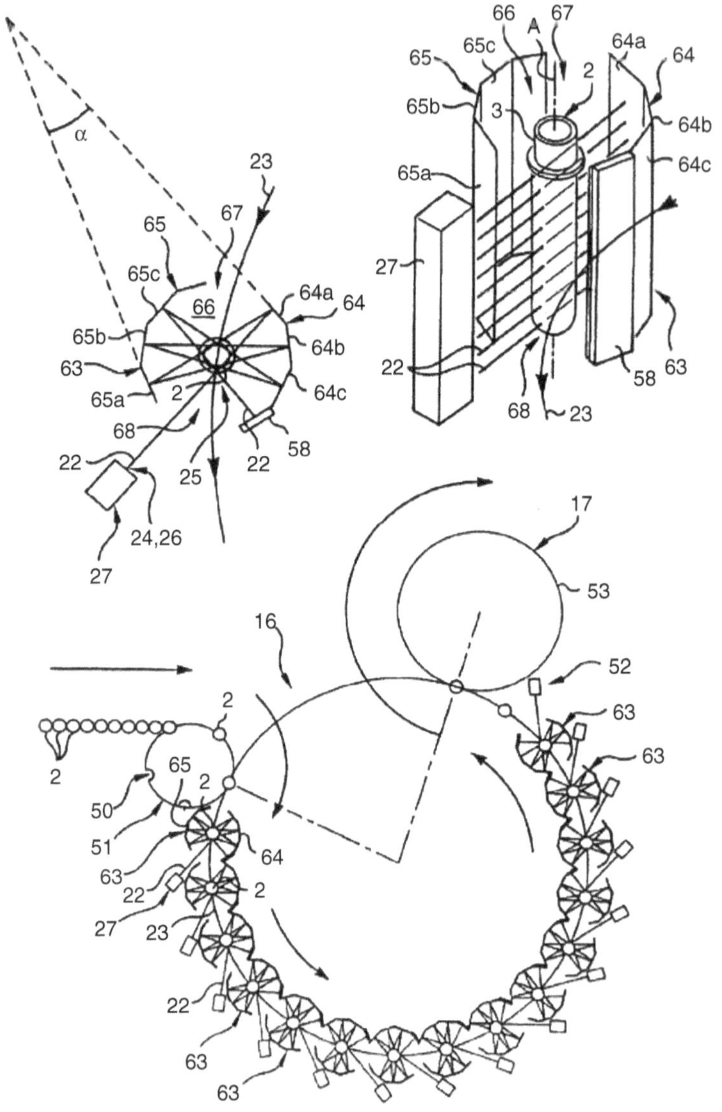

Figure 4.7 Circular oven with laser heating. Picture part of patent # 8,330,290.

control devices. Voltage regulators control all lamps from 0 to 220 V. On some control screens this is expressed as a percentage from 0 to 99%. Machines differ in how many voltage regulators are in use (Fig. 4.8).

In addition there is often a master module allowing the adjustment of all lamps in one oven up to the maximum voltage. Individual lamps can also be switched off completely either right on the screen or by pulling the

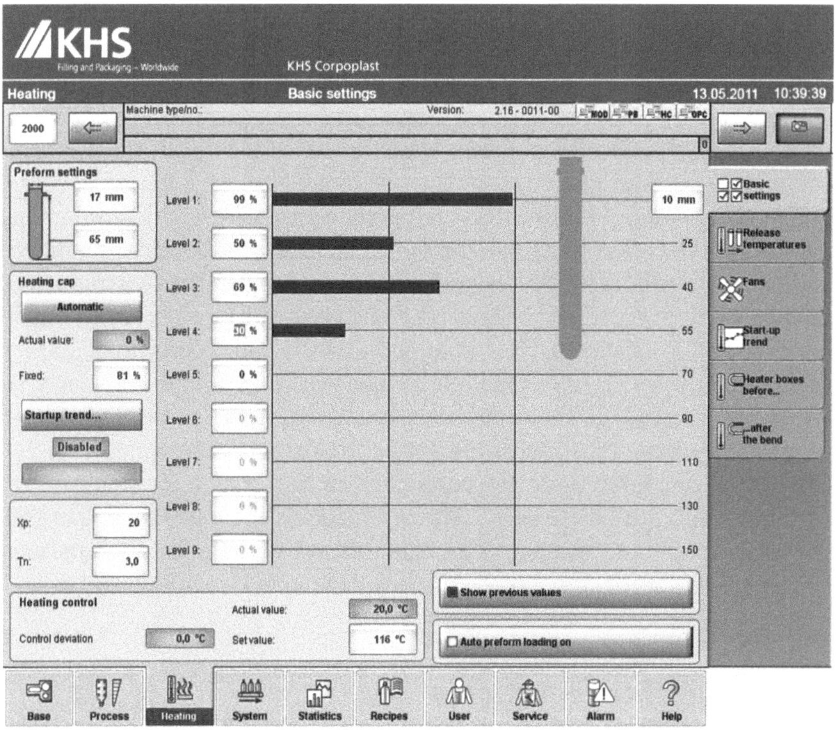

Figure 4.8 Modern machines offer lamp control right on screen. Diagram courtesy of KHS Corpoplast.

plug in the control cabinet (see Chapter 10 on how these features are used in the process).

Most machines monitor current flow to the lamps and warn operators if the current falls below a lower threshold because of lamp burnout or cable-breakage.

Fan Cooling

Besides cooling the lamps and reflectors oven systems must also provide preform cooling (Chapter 6). Designers use air blowers for this task exclusively and machines differ in the number, size, location, and control of these. Simpler methods are fans with fixed speed motors while more sophisticated machines offer variable speed motors to control air flow. In the latter case thermocouples in one oven provide a means of correlating the effect of the fan to the oven environment (see Chapter 6, Section 6.4 for possible problems). Close-looping of fan speed and oven temperature may or may not be available and is not always advised. Consider this scenario: Preform temperature is set to 90°C and oven temperature to 85°C, both being close looped. Let us say the plant temperature increases during the day as is common in the summer months in Northern climates. The effect on preform and oven temperature is not the same though. If the preforms become warmer the machine regulates the lamps down. This leads to a drop in oven temperature. If this drop is larger than the increased air temperature the fan will decrease its speed leading to increased preform temperatures and the lamps will be further decreased to compensate. The two regulatory circuits would basically fight each other unless the increase in environmental temperature change affects both parts by the same percentage, which could actually lead to overregulation when the lamp output drops and the fan speed increases at the same time. It is therefore often better to close-loop the preform temperature only and control oven temperature manually. Choose at least three times the number of cavities as the moving average for lamp control, that is, on a 20-cavity machine it should be at least 60. Choose a small proportional factor (the rate at which the lamp output will be changed) as well, typically 0.3. This will avoid overregulation (Fig. 4.9).

4.2 Transfer Functions

Rotary Machines

One variable speed DC or AC motor drives all machine functions through a combination of gears, chains, or linkages, ensuring that all physical movements are synchronized whether preforms are in the machine or not (except the stretch rod). Brakes and clutches on the main wheels prevent overloads or unwanted movements (spinning) during a sudden stop. An electronic rotary indicator signals the main drive position to the programmable logic controller (PLC) during each rotation. The PLC manages all nonsynchronized machine actions such as blow air.

A hand crank allows manual machine rotation for adjustment or clearing of preforms.

Transfer functions vary a great deal from machine to machine. Fig. 4.10 will give an idea of the components involved.

This machine keeps the preforms on the mandrels throughout the oven and blow section, turning bottles right side up after blowing. In this way the diagram differs from machines that handle preforms in other ways.

Preforms enter the machine at the loading station. Fed by gravity, worm drives, star wheels or scrolls, preforms are turned about 180 degrees when they go through the ovens neck down. Next they are pushed onto the

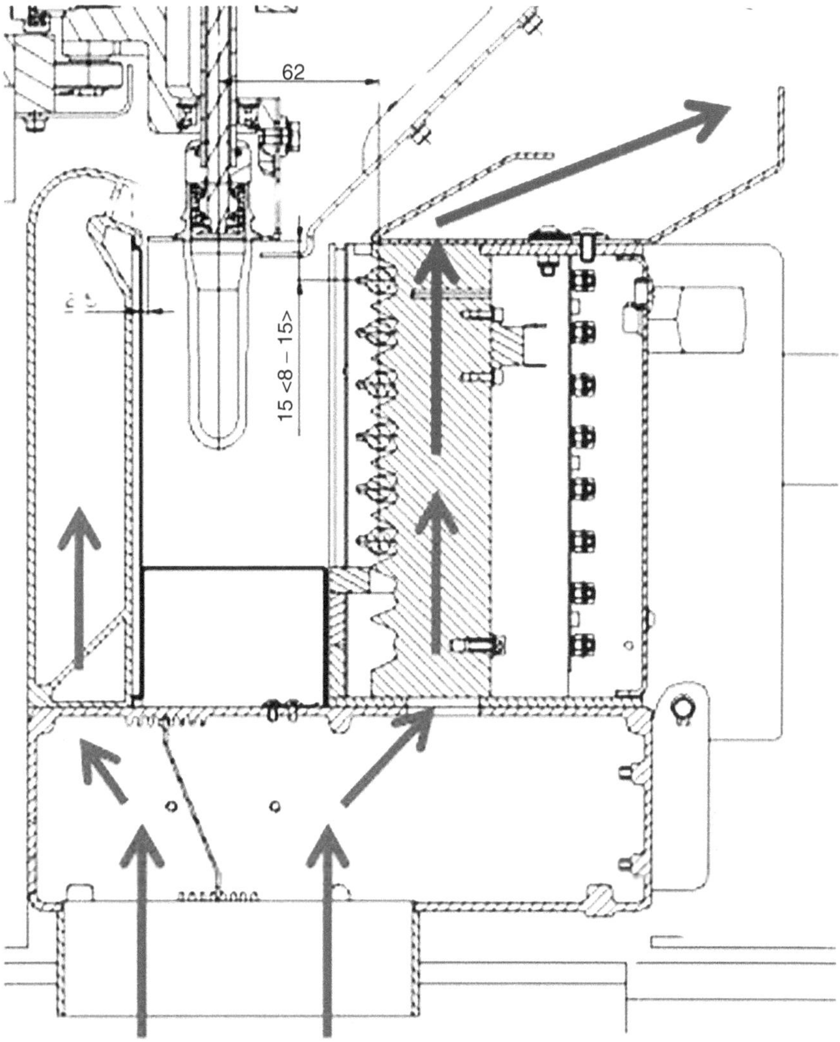

Figure 4.9 Air flow in oven with preforms right side up. Diagram courtesy of Adphos Group.

mandrels, which is accomplished by either holding them in nests and moving the mandrels up from underneath or holding the mandrels and pushing the preforms into place. All machines have devices ensuring the preforms are firmly seated on the mandrels. Preforms that are not located firmly are removed from the machine by various means (Fig. 4.11).

The loading/unloading wheel turns clockwise as do the heating and blow wheels. The intermediate or transfer wheels turn counterclockwise.

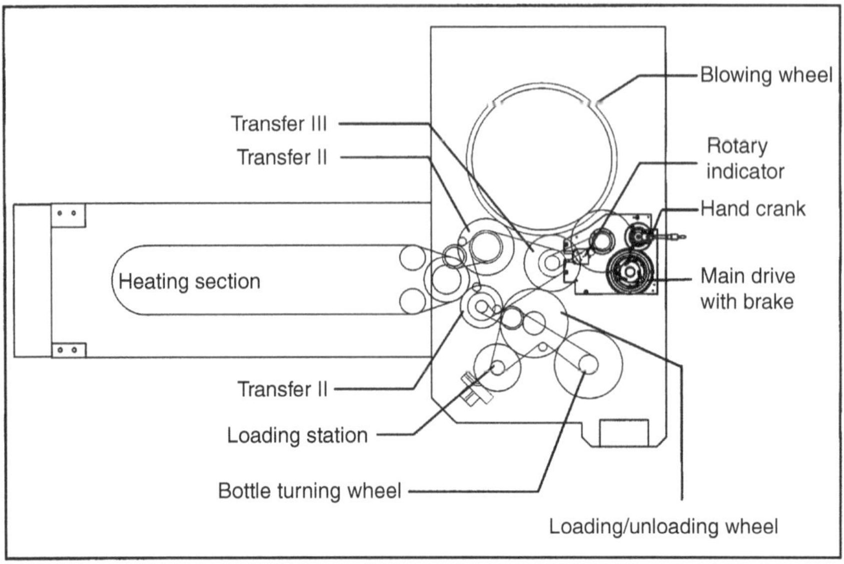

Figure 4.10 Layout of the various transfer functions of a Blomax Series III blow molding machine. Diagram courtesy of SIG Corpoplast.

Figure 4.11 Preforms not properly seated on the mandrels are blown into an ejection pipe. Photo courtesy of SIG Corpoplast.

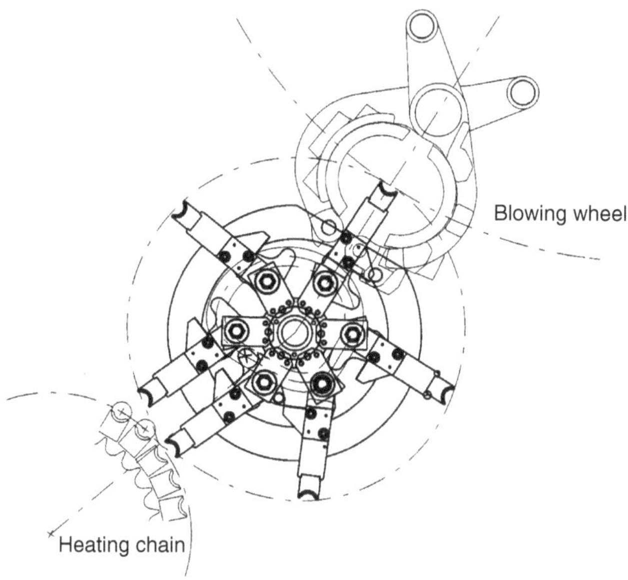

Figure 4.12 The loading/unloading wheel is the heart of the various preform and bottle transfers. Diagram courtesy of SIG Corpoplast.

Loading of preforms onto mandrels and transferring blown bottles from transfer III to the turning wheel are both accomplished by this first wheel. This dual functionality, that is found throughout rotary blow molding machines, is possible because the neck support rings, that are already fully finished in the preforms, are used to hold both preforms and bottles (Fig. 4.12).

Transfer I takes mandrels with preforms to the heating chain from where they travel through the oven section. Transfer II then moves them from the heating chain to the blow wheel (Fig. 4.13).

The elongated shape of the transfer arms or grippers is due to the fact that they must take the preforms to and from the center of the blow clamp, clearing the mold halves.

Transfer III finally takes the blown bottles from the blow clamps to the loading/unloading wheel. Because they are still upside down on this machine, they are then transferred one more time into the bottle turning wheel before they exit the machine.

Linear Machines

These machines often have devices that turn a number of preforms and bottles through 180 degrees. This number is governed by the number of cavities that the machine blows (Fig. 4.14).

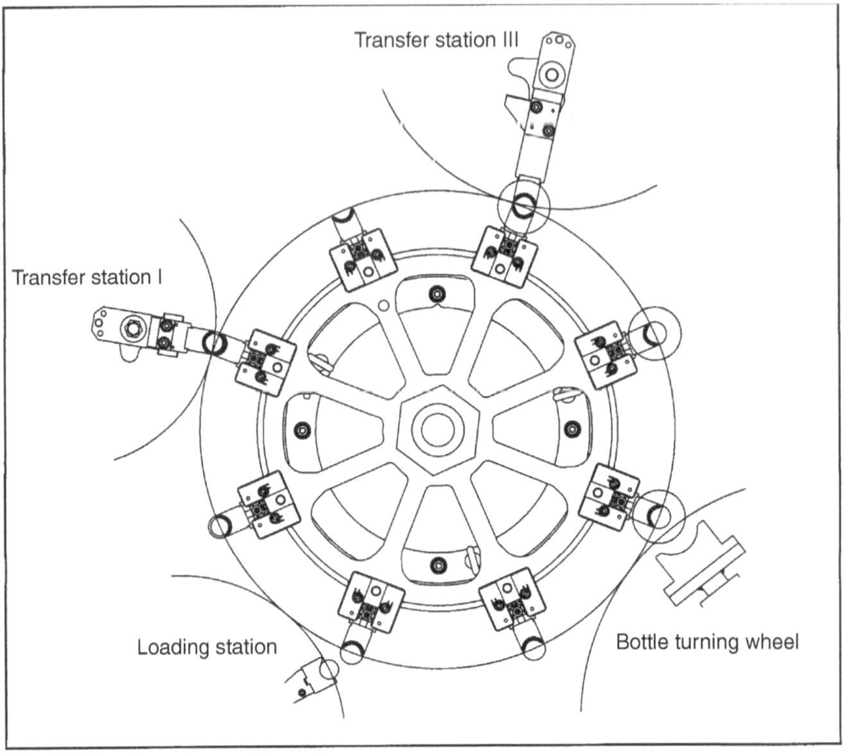

Figure 4.13 Transfer wheels transfer preforms and bottles from one machine function to the other. Diagram courtesy of SIG Corpoplast.

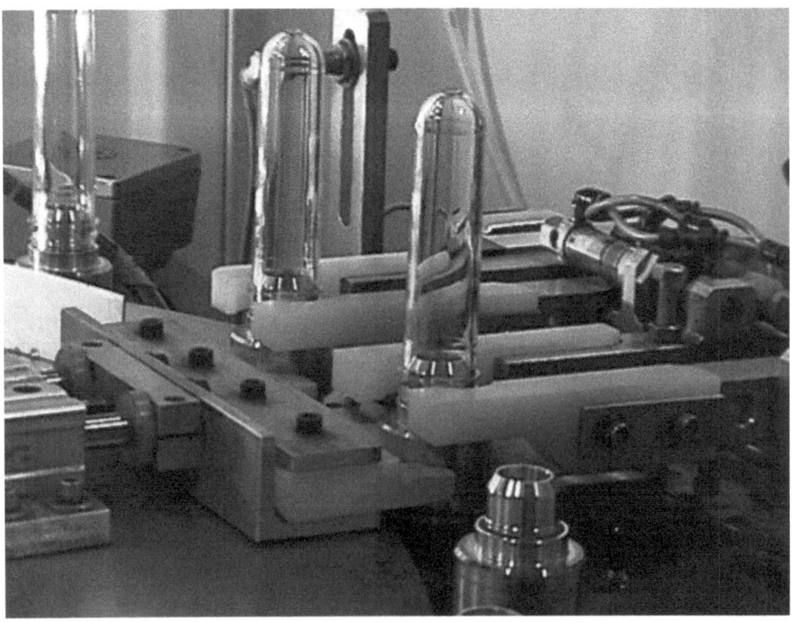

Figure 4.14 The cylinder on the left centers and holds the mandrels while the preform turning device places two preforms. Photo courtesy of Amsler Equipment.

Most machines use individual mandrels, combining them in front of the blow clamp and shuttling them in with a hydraulic cylinder. Mandrels may wear where they have hardened locating rings that fit into cut-outs in the blow mold halves.

4.3 Blow Wheel/Blow Clamp

Rotary Machines

Blow clamps are highly engineered, complex devices contributing significantly to the rather large price premiums that rotary machines demand when compared with linear machines of the same cavitation. Each clamp has its own valves, water, and electrical connections supplied from the center of the blow wheel.

As mentioned before cams control all blow mold movements. While the blow wheel turns carrying the blow clamps, stationary cams engage through rollers and levers guiding the linkages into the desired movements.

Moveable mechanisms deflect cams and levers in the case of a malfunction.

Blow clamps are built to accept the most common bottle sizes ranging from about 200 mL to 2.5 or 3 L in volume. There are also minimum and maximum preform and bottle dimensions that must be respected.

Over the years engineers have significantly lightweighted blow clamp components. This in turn has allowed the increase in throughput of the machines. To achieve this the closing mechanisms do not themselves hold the molds closed, which would require them to resist the large forces created by the high pressure. Instead, locking mechanisms of various designs engage after mold closing. To reduce the stress on these, blow air in small quantities acts on the outside of the mold thus balancing the forces inside and outside of the blow clamp (Figs. 4.15 and 4.16).

4.4 Machine Timing

In rotary machines the control of the various machine functions is facilitated by using the 360 degrees of the turning wheel as trigger points rather than timers. Typically, timing starts when the mold is fully open. Each movement then happens with a start and end degree. For example, the mold may close between 14 and 44 degrees. How long this actually is depends on the machine speed. Table 4.2 lists the converting times in degrees.

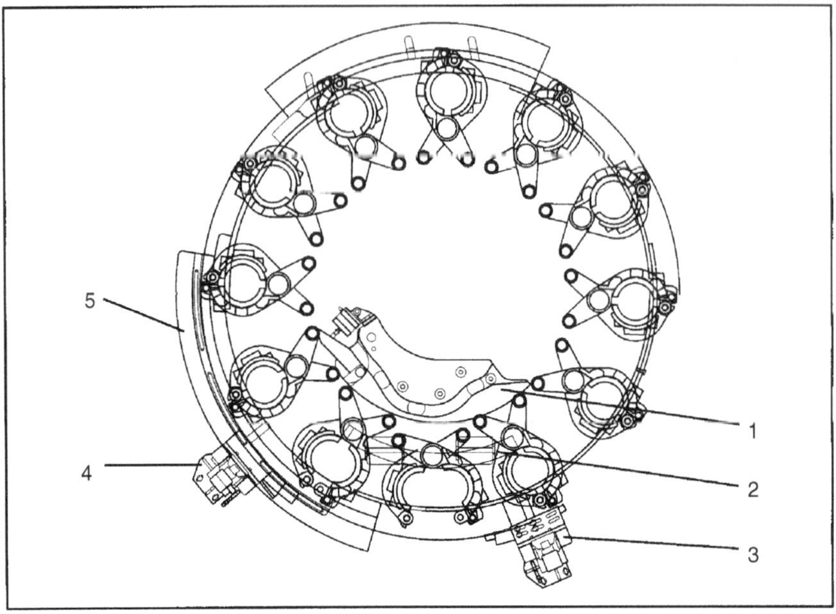

Figure 4.15 Drawing numbers and their respective functions: *1*. cam "opening/closing of the mold"; *2*. cam "lifting the base mold"; *3*. cam "locking the mold"; *4*. cam "unlocking the mold"; *5*. cam "lifting/lowering the stretch rod."

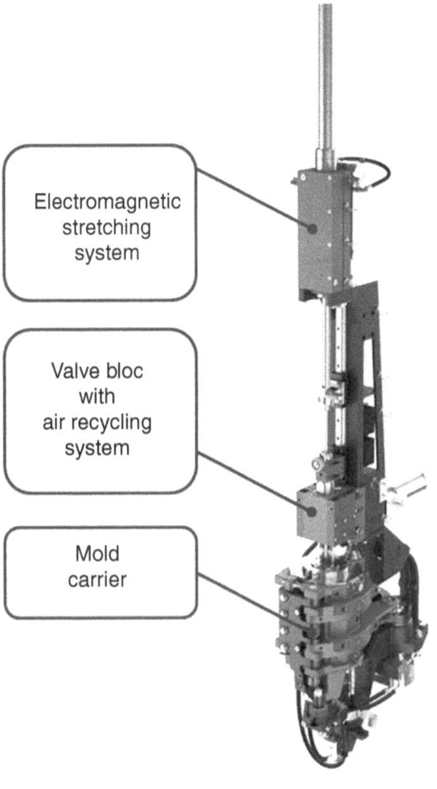

Figure 4.16 Blow clamp and components. Diagram courtesy of Krones AG.

Table 4.2 **Depending on the Speed of the Machine the Same Time Leads to Different Angle Displacements**

Time	Bottles per Cavities per Hour (Degree)							
(s)	**1500**	**1600**	**1700**	**1800**	**1900**	**2000**	**2100**	**2200**
0.1	15	16	17	18	19	20	21	22
0.15	22.5	24	25.5	27	28.5	30	31.5	33
0.2	30	32	34	36	38	40	42	44
0.25	37.5	40	42.5	45	47.5	50	52.5	55
0.3	45	48	51	54	57	60	63	66
0.35	52.5	56	59.5	63	66.5	70	73.5	77
0.4	60	64	68	72	76	80	84	88
0.45	67.5	72	76.5	81	85.5	90	94.5	99
0.5	75	80	85	90	95	100	105	110
0.55	82.5	88	93.5	99	104.5	110	115.5	121
0.6	90	96	102	108	114	120	126	132
0.65	97.5	104	110.5	117	123.5	130	136.5	143
0.7	105	112	119	126	133	140	147	154
0.75	112.5	120	127.5	135	142.5	150	157.5	165
0.8	120	128	136	144	152	160	168	176
0.85	127.5	136	144.5	153	161.5	170	178.5	187
0.9	135	144	153	162	171	180	189	198
0.95	142.5	152	161.5	171	180.5	190	199.5	209
1	150	160	170	180	190	200	210	220

Table 4.3 is the reverse table, converting degrees into times.

This now has the peculiar effect that process time cannot be simply calculated. Here is an example to demonstrate the following.

Assuming that a machine runs at 2200 b/c per h for a lightweight water bottle and the mold closing takes 30 degrees. It would therefore take 0.136 s to complete. When the machine is then slowed down to 1500 b/c per h to make a thicker-walled bottle the same movement now takes 0.2 s. In most rotary machines process time (preblow, blow, and exhaust) is about 70% with 30% necessary to complete the mechanical machine functions. A 30% dead time or unproductive time translates into 108 degrees and at 2200 b/c per h into 0.49 s whereas at 1500 b/c per h it is already up to 0.72 s or an almost 50% increase, although small in real numbers.

Most linear machines use a higher percentage of overall time for unproductive time as it takes longer to shuttle the preforms and bottles in and out of the mold, and open and close of the mold. However, there is no change

Table 4.3 Depending on the Machine Speed Different Degrees of Machine Rotation Take Different Times

	Bottles per Cavities per Hour							
Degrees	**1500**	**1600**	**1700**	**1800**	**1900**	**2000**	**2100**	**2200**
15	0.100	0.094	0.088	0.083	0.079	0.075	0.071	0.068
30	0.200	0.188	0.176	0.167	0.158	0.150	0.143	0.136
45	0.300	0.281	0.265	0.250	0.237	0.225	0.214	0.205
60	0.400	0.375	0.353	0.333	0.316	0.300	0.286	0.273
75	0.500	0.469	0.441	0.417	0.395	0.375	0.357	0.341
90	0.600	0.563	0.529	0.500	0.474	0.450	0.429	0.409
105	0.700	0.656	0.618	0.583	0.553	0.525	0.500	0.477
120	0.800	0.750	0.706	0.667	0.632	0.600	0.571	0.545
135	0.900	0.844	0.794	0.750	0.711	0.675	0.643	0.614
150	1.000	0.938	0.882	0.833	0.789	0.750	0.714	0.682
165	1.100	1.031	0.971	0.917	0.868	0.825	0.786	0.750
180	1.200	1.125	1.059	1.000	0.947	0.900	0.857	0.818
195	1.300	1.219	1.147	1.083	1.026	0.975	0.929	0.886
210	1.400	1.313	1.235	1.167	1.105	1.050	1.000	0.955
225	1.500	1.406	1.324	1.250	1.184	1.125	1.071	1.023
240	1.600	1.500	1.412	1.333	1.263	1.200	1.143	1.091
255	1.700	1.594	1.500	1.417	1.342	1.275	1.214	1.159
270	1.800	1.688	1.588	1.500	1.421	1.350	1.286	1.227
285	1.900	1.781	1.676	1.583	1.500	1.425	1.357	1.295
300	2.000	1.875	1.765	1.667	1.579	1.500	1.429	1.364
315	2.100	1.969	1.853	1.750	1.658	1.575	1.500	1.432
330	2.200	2.063	1.941	1.833	1.737	1.650	1.571	1.500
345	2.300	2.156	2.029	1.917	1.816	1.725	1.643	1.568
360	2.400	2.250	2.118	2.000	1.895	1.800	1.714	1.636

in this whether the machine runs at maximum speed or not; these movements always take the same amount of time.

Potential buyers should always ask what the process time is for a given output and then judge from their range of applications which type of machine fits best.

4.5 Rotary Machines Comparison

While there are many suppliers of linear machines, the rotary machine market is dominated by four suppliers:

Table 4.4 An Overview of the Various Parameters of Rotary Machines

Parameter	KHS	Krones	Sidel	SIPA
Max. b/c per h (small water)	2,250	2,250	2,300	2,250
Max. bottle size (L)	3	3.5	3.5	3
Cavitation 0.7 L max.	4–36	8–36	6–34	6–24
Maximum transfer ring	36/48	38/48	38/48.5	38/43
Preform orientation	Up	Up	Down/up	Down/up
Stretch rod movement	Servo	Electromagnetic	Cam/air	Cam/air
Max. output (b/h)	81,000	81,000	78,200	50,400

- KHS Corpoplast, Germany (Blowmax series IV)
- Krones, Germany (Contiform Generation 3)
- Sidel, France (Matrix)
- SIPA, Italy

Since this market produces the vast percentage of all PET bottles, it is worth to compare the different machine characteristics so the reader can better understand what is different about them. I will examine the latest offerings in Table 4.4.

One should note that only bottles with stretched-out bases, such as lightweight water bottles, can be molded at these speeds. Thicker bases require longer cooling time to avoid the creeping out of the base center, which can lead to an additional "leg" causing the bottle to rock (the so-called rocker bottom). In carbonated soft drinks (CSD) bottle production, this is especially important as the center disk is stressed significantly and must stay rather thick. This in turn is required by the pressure cycles inside the bottle that may reach up to 5 bar (70 psi) when the bottle contents are agitated as happens during a bumpy truck ride. One of the parameters that quality control has to monitor is the so called base clearance to make sure this does not happen.

All machines show decreased output as the number of cavities increases. This is because the centrifugal forces become rather large with the larger blow wheel diameters.

The trend in blow molding machines has been to eliminate the turning of preforms and/or mandrels, which reduces handling and wear issues that have plagued some of the systems. Only Sidel turns the preforms after the oven section up, whereas SIPA uses what is referred to as a "caterpillar" oven where preforms travel upward and downward on their mandrels but there is no actual turning.

Figure 4.17 Instead of opening sideways, this clamp opens one side vertically. Picture courtesy of SIPA.

Both Sidel and Krones use standard clamping devices with both mold sides opening. KHS has moved to one stationary and one moving side, reducing blow-wheel footprint. SIPA uses what they call the "crocodile" mold movement, with one side being stationary while the other opens to the front instead of to the side (Fig. 4.17). It is also a space-saving measure.

Changeover Times

Changeover times are important for those factories that do not have the luxury of running the same product 365 days/year but have to change 2–5 times per week. When the same necks are used, the steps involved are as follows:

- disconnect water hoses
- replace blow molds

Figure 4.18 This tool-less system allows mold changeover in less than 1 min/cavity. Picture courtesy of Sidel Inc.

- replace bottom inserts
- replace stretch rod stops or dial in new stop values
- change PLC recipe
- change preform supply.

All blow machine manufacturers have now quick-change systems (Fig. 4.18 that make this task much less time consuming.

- Water hoses are equipped with quick-disconnects or eliminate the disconnection completely by feeding water through the carriers with the molds being outfitted with O-rings that do not require anything to be disconnected.
- Blow molds used are mounted with two screws holding each half. Modern systems feature locking mechanisms that allow mold removal with one twist of a special tool or no tool at all.
- Same for the bottom inserts.
- Stretch rod stops are also easily replaced or not necessary when servomotors are used instead of cam-driven pneumatic cylinders.

Taken all measures together may reduce changeover time from 5 min/cavity to less than 1 min/cavity. On a 20-cavity system, this would result in time savings of 80 min/changeover, allowing an additional production of over 53,000 bottles if the machine runs at a speed of 2,000 bottles/cavity per h (b/c per h). Even if the margin was only 0.5 cents/bottle, an additional revenue of $250 per changeover could be achieved.

Air Consumption

Another important feature is air consumption and more specifically the so-called dead air volume. Air is the single-most costly part of blowing bottles besides the preforms and users do well to pay close attention to this issue.

Air consumption is driven by the air pressure, the actual volume of the bottles produced, and the air that is not used in the blow process but needs to be exhausted every blow cycle. This air "hides" in the connection pieces between the blow valves and the cavity. When the exhaust valve is energized, all air that is between it and the cavity must be exhausted besides the actual bottle volume to make sure there is no pressure left in the blown bottle. Many manufacturers of blow machines do not publish this important information but users can calculate it by looking at the machine specifications.

Let us assume a manufacturer specifies blow air consumption of 1300 Nm^3/h for the production of 20,000 b/h of 1.5 L volume at 35 bar. The volume that is actually inside the bottles is 20,000 b/h $\times$ 1.5 L $\times$ 35 bar/1,000 L/m^3 = 1,050 Nm^3/h. The difference of 1300 − 1050 = 250 Nm^3/h is the dead air volume. This in turn is 367 mL/bottle (250 Nm^3/h $\times$ 1000 L/Nm^3/h/20,000 b/h/35 bar). It is important to calculate this as a fixed number rather than a percentage as this lost air is present for bottles of each size. A 500-mL bottle has the same losses as a 1.5-L bottle; however, in the case of the smaller bottle 73% more blow air is needed than would be if just the bottle needed to be supplied. Smaller dead air volumes will save significant money over the lifetime of the machine. Small machines particularly often feature high dead air losses and users should carefully examine manufacturers' offerings (Fig. 4.19).

There is one more source of needed air volume that is not mentioned yet. On some machines that use pneumatic cylinders for stretch rod activation, this air is actually high-pressure blow air that is reduced to the 8 bar or so that the stretch rod cylinders are using. While compressing air to 35 or 40 bar and then reducing it to 8 or 10 bar is an expensive undertaking, this is done to avoid adding another line to the distributor inside the blow machine as this distributor for air, water, and electrical power is one of the

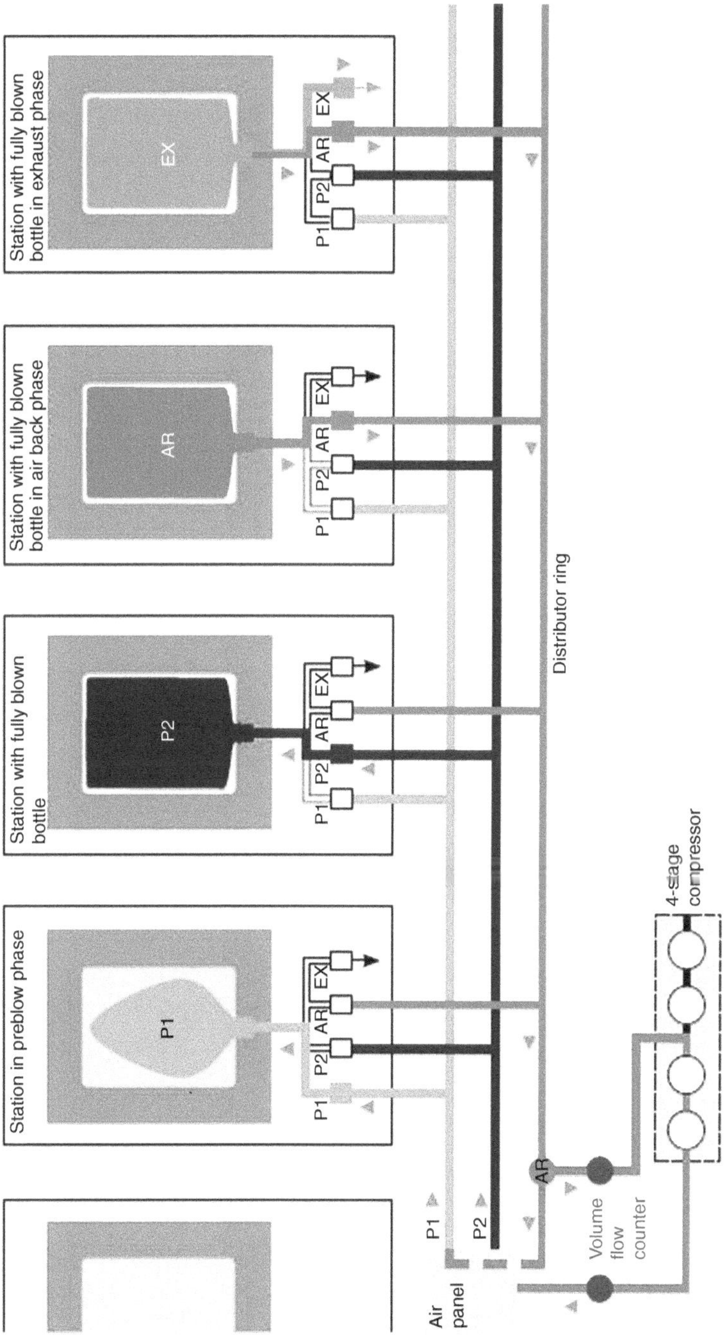

Figure 4.19 During the "airback" phase, blow air is redirected to a distributor panel instead of exhausted to the atmosphere. Diagram courtesy of KHS Corpoplast GmbH.

most expensive parts of the blow molding machine. The newest machines with servo or electromechanically controlled stretch rods do of course not use any air and will save money in the long run.

Because of the cost of air consumption, most companies now offer air-recycling systems. These systems, instead of exhausting blow air into the atmosphere, pipe it back to a storage tank where it can then be used for preblow and stretch rod air, the machine air circuit, or other plant air requirements. Of course, pressure reducers in the systems allow the proper pressure for the various uses. In this manner, savings of 25–50% can be achieved. This not only reduces operational costs but also helps with capital expenses as fewer or smaller compressors may be needed. A small cycle time penalty is unavoidable because the air does not move as fast to the pressurized storage tank as it does to the lower atmospheric pressure.

5 Blow Molds

Chapter Outline

Blow molds play a large part in making high-quality bottles. While the machine has to deliver preforms at the right temperature, it is the blow molds that give containers repeatable features and a brilliant appearance (Fig. 5.1).

5.1 Design

Neck and thread finish are already formed in the preform; so blow molds form only the body and base of the bottle. In the reheat stretch blow molding (RSBM) process they consist of three parts: two mold halves and one base insert (also called push-up). The base insert is necessary because the walls at the base of the concave container could not slide over the mold halves during mold opening if these were forming them. Instead, the vertically moving base insert is drawn out of the way before, or as, the mold opens (Fig. 5.2).

While the three-piece design is common to all molds, they are manufactured quite differently depending on the type of machine to which they will be fitted. Linear machines have all mold cavities mounted within two blocks where the cavities sit side by side. In rotary machines each blow mold is mounted to a separate carrier, opening and closing individually. Modern machines use so-called shell molds whereby the actual mold halves are only 5-mm thick and are assembled onto bases that are all the same for a family of containers. These bases carry all water connections and need not be touched during a changeover, thus reducing valuable time.

Molds are usually built from aluminum, which is chosen for its high heat transfer rate, easy machinability, and lightweight. The types of aluminum used are typically those used in the aircraft industry. AL 7075 T6, T-2024, or Alumenec 89 are some of the grades used worldwide for this application. Base inserts may be of the same material or made from beryllium–copper.

Stretch Blow Molding. http://dx.doi.org/10.1016/B978-0-323-46177-1.00005-6

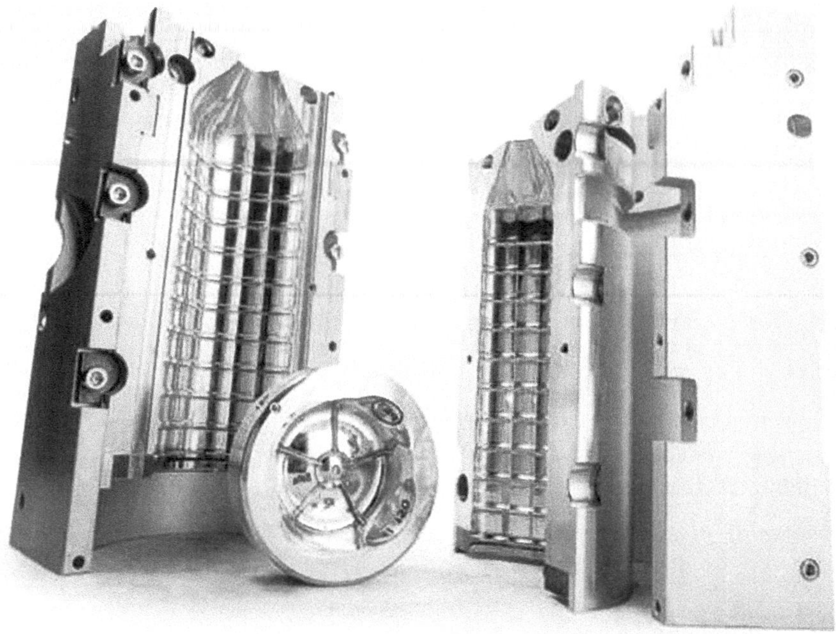

Figure 5.1 Blow mold halves in and out of mold base and base insert. Highly polished aluminum is the most commonly used material for standard bottles. Photo courtesy of SIPA.

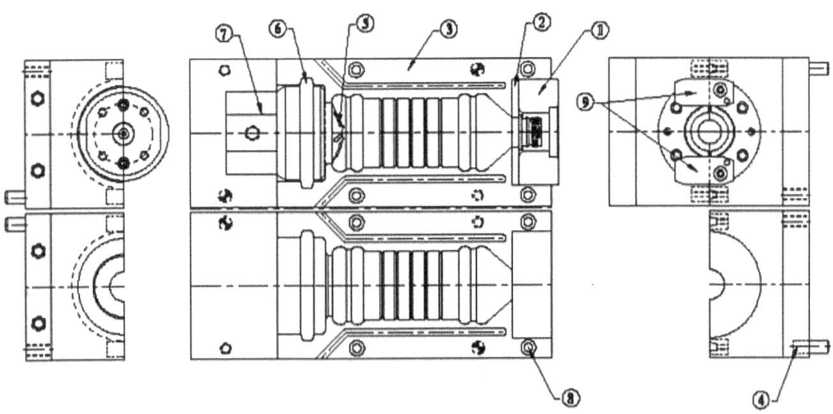

Figure 5.2 Typical mold design for linear RSBM machine. The given details are followed by their descriptions: *1*, Preform Retainer Insert; *2*, S.S. Insert; *3*, Mold Body; *4*, Back Plate; *5*, Base Insert/Push-up; *6*, Locating Ring; *7*, Push-up Holder; *8*, Taper Lock Pins and Bushings; *9*, Guide Finger. Drawing courtesy of Hallink Molds Ltd.

Another material in use is stainless steel for hot-fill applications. While a polish to mirror-like quality is still the norm, some companies have quite successfully tried to leave molds at a much rougher polishing state. This saves cost because the mirror polish is still applied manually while detracting only very slightly from the expected bottle surface appearance.

The internal pressure of the blow air results in a considerable force against the closing mechanism of the blow mold. Blow molds have guide pins and bushings as well as taper locks in the base insert that keep the mold in position during the blow process. Mold carriers feature locking mechanisms that keep them closed against the blow pressure. To alleviate these stresses and also make ever-lighter frames possible, "pancake" cylinders have become increasingly popular. These cylinders are very thin shells behind the blow molds and are filled with the same air that blows the bottle. Since pressure is equal both inside and outside the mold there is no resultant force acting against the mold halves (Fig. 5.3).

Figure 5.3 Blow mold with locking mechanism mounted inside a rotary machine. Photo courtesy of SIPA.

5.2 Base Mold

Today, all base molds feature a small recess like a well in the center. This allows some room for the protruding injection gate on the preform. Keeping the gate in the center of the mold is probably the most important task in the blow process because any deviation from the center leads to uneven wall thickness variation. The well catches the injection gate and prevents it from slipping as long as there is enough pressure from the stretch rod.

In many custom applications the bottle bottom is thicker than is really needed but because of preform design or machine insufficiencies it ends up like that. In these cases it is often the cooling of the bottle bottom that controls the cycle time. Proper cooling is therefore crucial to come to a cost-effective solution with fast cycles. Cooling lines should not be smaller than 6 mm in diameter (unless of course there is no room in very small bottles) and the flow path should have no restrictions. High water supply pressure with low water return pressure are also helpful (Chapter 13, Section 13.2).

5.3 Making a Mold

Today's mold-making process starts with a three-dimensional (3D) computer model of the container itself. Physical models may be made by a variety of processes, the most popular still being stereo lithography with 3D printing catching up quickly because of the availability of low-cost printers. The model may be used to give marketing people a better "feeling" for a new container. Once approved, data of the computer model are then fitted in a new or existing mold base. At this point, shrinkage has to be added to the container dimensions. Polyethylene terephthalate (PET) shrinks approximately 0.08% but shrinkage is not uniform and it is the experience of the mold maker that determines how closely the capacity of the container matches specification.

A variety of computer-aided design (CAD)/ computer-aided manufacture (CAM) programs allow the creation of machine cutter paths that are downloaded directly into high-speed machining centers. Machine operators load and center blocks of aluminum of suitable size and special cutters, spinning at up to 30,000 r.p.m., move at a speed of up to 20 m/min. The resulting cavity surface is already smooth to the eye but most mold makers add a high, mirror-like polish, which still requires skilled, manual labor. The use of sandblasted surfaces that are common in other plastic processes has gained some ground as there is little difference in the appearance of the containers. Some mold makers then coat the cavity surfaces with various materials, often containing nickel and Teflon, to give it abrasion resistance.

5.4 Venting

Venting is another area where the experience of the mold maker becomes extremely important. Because PET fills the mold cavity during blowing, the air inside the cavity must be exhausted. For this purpose mold makers add a variety of vents. Compared to other processes, such as injection molding or extrusion blow molding, PET is processed at a relatively low temperature in the RSBM process. Vent sizes are limited to 0.04 mm (0.0015 in.) in injection molding but vents of up to 0.5 mm (0.020 in.) are used in RSBM with hole vents up to 1 mm (0.040 in.). All molds have vents on the contact surface of the cavities. One mold half is typically completely recessed against the mold base by up to 0.20 mm (0.08 in.) or more commonly by 0.15 mm (0.006 in.). Base vents are also common and are accomplished by leaving the base insert to move 0.25–0.3 mm (0.010–0.012 in.) downward under the force of the stretch rod. The resulting ring-shaped gap between base insert and mold cavity allows air to escape.

Hole vents up to 1 mm are used in areas where air entrapment is suspected. Vents of this diameter may not show in areas where the material has stretched and consequentially strain-hardened but will show as small dimples where this is not the case. A common example of highly stretched material is the foot of a petaloid base for carbonated soft drinks (CSD) containers. Two small holes in each foot let air escape that might otherwise be trapped by the material flowing around it (Fig. 5.4).

Another use of venting is to direct PET into hard-to-blow areas. In a highly oval bottle, for example, there is always the possibility of a ridge of higher wall thickness forming at the center of the narrow side of the container. Vent holes at the far side of the mold can attract PET to flow more quickly into these areas, thereby stretching out the preform walls close to the narrow side. A fine sandblast finish instead of the mirror-finish also helps to let the air move out of the mold.

Due to low temperature in the RSBM process compared with uses in other processes, PET does not flow easily into small mold crevices. Minimum dimensions for female radii might be given as 0.8 mm

Figure 5.4 Hole vents up to 1 mm in diameter can be successfully used as shown here in the panel area of a hot-fill bottle. Photo courtesy of Garrtech Inc.

(1/32 in.) but it will depend on the stretch ratio of the PET flowing toward it whether it will fill out or form a greater radius instead. Male radii should be double that amount especially when used in bases. Here a sharp radius may cause a crease in the material and open the door to stress cracking. Venting in these areas can be attempted to reduce the risk of air entrapment stopping the advance of the parison but more often than not they do not seem to have much effect. We will simply have to live with the fact that PET benefits from more generous radii in this process.

5.5 Stretch Rod

In two-stage molding stretch rods are typically made of solid Stainless Steel with sizes 9–16 mm (3/8–5/8 in.). Stretch rod diameters are often limited by the neck size and larger rods should always be used when the neck size allows it. This facilitates the exact placing and holding of the gate in the center of the mold. A downside of a large stretch rod diameter is that the surface of the rod cools the area it touches on the preform. Lightweight water bottles, for example, need the base totally stretched and a large rod diameter may prevent that from happening. On the other side, making a rod too thin would lend it to easy bending. As a consequence, rods are made to keep the larger diameter in the upper portion

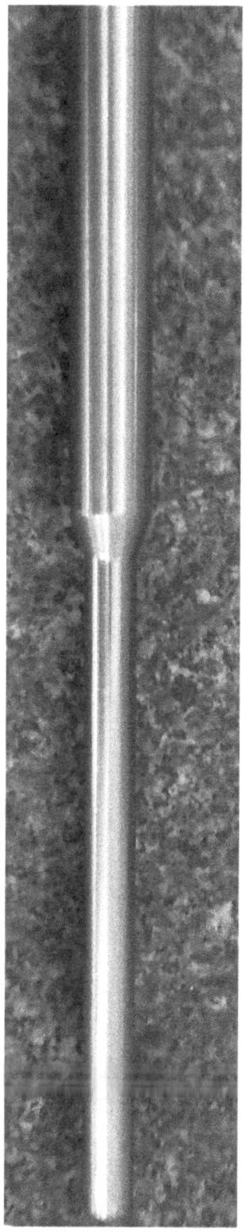

Figure 5.5 For lightweight water bottles and some soft drinks bottles stretch rods are slimmed down to prevent a cooling effect on the preform bottom.

to prevent bending but are reduced in diameter in the area of the preform (Fig. 5.5).

In single-stage molding, stretch rods often have tips screwed on at the end. These could be made from aluminum or nylon. On machines without conditioning, they can also be used to cool the bottom of the preform, just the opposite function they have in two-stage molding. This is useful because the bottom of the preform is often the hottest part and may lead

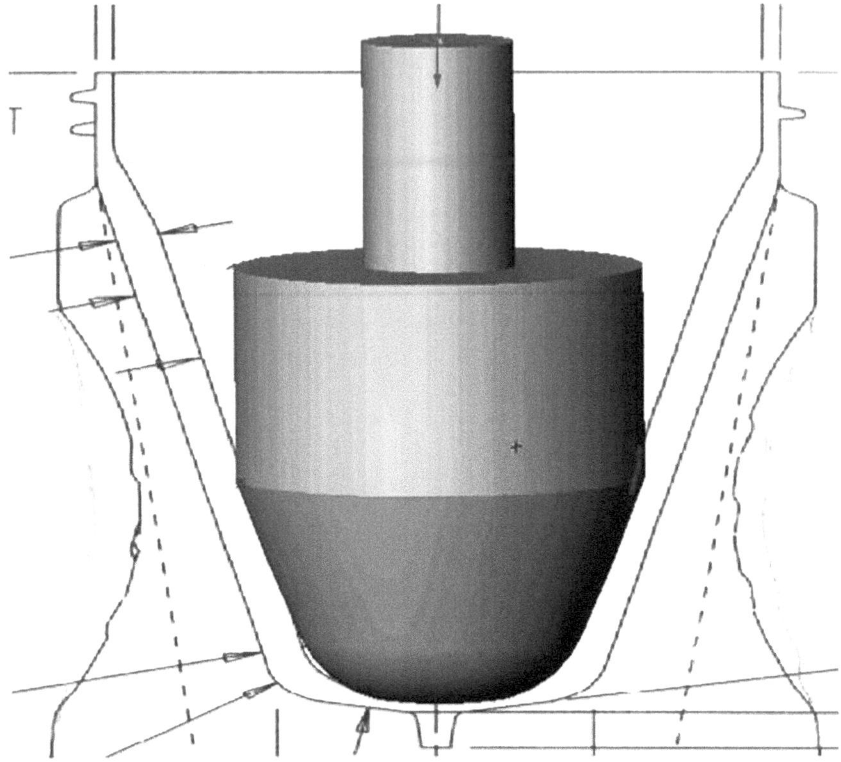

Figure 5.6 In single-stage molding a large stretch rod tip can help in cooling the bottom of the preform down to increase wall thickness in the bottom of the bottle.

to thin bottoms. A large tip made from aluminum can be placed into the preform a few seconds before the mold is closing. This does not lead to a cycle time increase as the blow cycle is always faster than the injection cycle. The timer "delay mold close" is used to achieve this effect (Fig. 5.6).

6 Fundamentals of the Blow Process

Chapter Outline

6.1 Process Overview

While individual features differ between different types and brands of machines, they all load cooled, injection-molded preforms, reheat them in an oven section, transfer them in a blow mold clamp, and stretch and blow them (for differences to the single-stage process see Chapter 8) (Fig. 6.1).

Several questions arise that will be answered in this section of the book:

- What is the optimal blowing temperature?
- What is the optimal heat profile in the preform wall?
- How does polyethylene terephthalate (PET) absorb infrared energy and how can this knowledge be used to improve bottle properties?
- How do material properties and stretch ratios affect blowing behavior?
- How can optimal wall thickness distribution be achieved?
- What amount of oven cooling should be used?
- How does blow air timing affect bottle properties?

These are questions that machine operators are faced with daily and often a compromise between machine capabilities, output requirements, and bottle properties has to be found. It is therefore paramount that processors understand all relevant process characteristics. Only then are they able to find the optimal solution to a multifaceted problem.

6.2 Stretch Ratios

To better understand what the process needs to accomplish, we need to study how stretch ratios between bottle and preform are calculated and how temperature differences in the preform wall affect them. While there is no

Stretch Blow Molding. http://dx.doi.org/10.1016/B978-0-323-46177-1.00006-8

Figure 6.1 Standard preform for carbonated beverages. The ring underneath the thread, called neck support ring (NSR) or transition ring, is used throughout the blow molding machine to hold the preform in place.

accepted standard in the industry how stretch ratios are calculated, it is safe to say that they represent the ratio of bottle dimensions to corresponding preform dimensions. There are two types of stretch ratios (Fig. 6.2):

- Axial stretch ratio dividing the length of the bottle by the length of the preform as measured from underneath the neck support ring (NSR) to the end of bottle and preform.
- Hoop stretch ratios can be measured in three different ways:
  - Dividing the outside bottle diameter by the corresponding outside preform diameter.
  - Dividing the inside bottle diameter by the corresponding inside preform diameter.
  - Dividing the average bottle diameter by the corresponding average preform diameter.

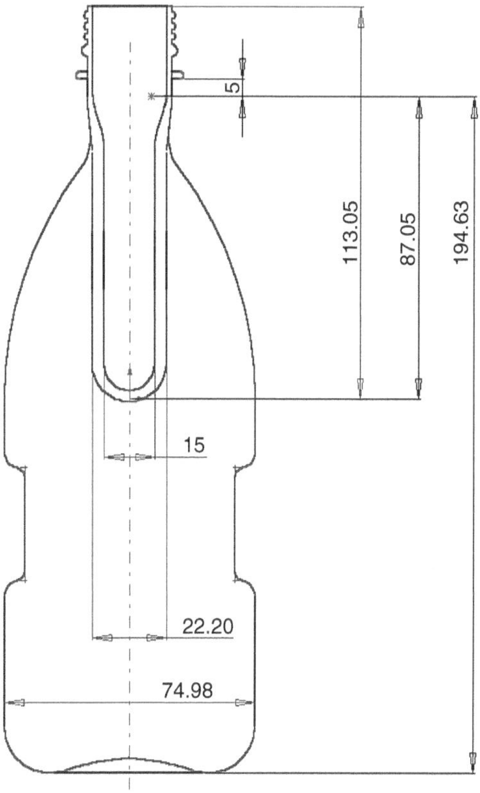

Figure 6.2 Stretch ratios can be calculated from preform and bottle dimensions.

Taking Fig. 6.2 as an example, here are three calculated stretch ratios:

Outside hoop stretch ratio (max.):	74.98/22.2 = 3.38
Inside hoop stretch ratio (max.):	74.58/15 = 4.97
Axial stretch ratio:	199.63/92.05 = 2.17

It becomes apparent that the inside stretch ratio is significantly (47% in this particular case, others may be lower or higher depending on the wall thickness of the preform) higher than the outside stretch ratio. This is the result of the bottle wall thickness being so much smaller than the preform wall thickness, so there is only a small difference between inner and outer bottle diameter but a large difference between inner and outer preform diameter. Intuitively, it is understood that warmer areas stretch easier than colder ones.

Since the inner preform walls must stretch further, we can postulate that they should be at a higher temperature than the outside walls. However, heat enters the preform from the outside by means of the infrared lamps, so it seems a rather difficult challenge to get the inside wall to a higher temperature. In order to understand how this can indeed be accomplished, we need to understand how the oven lamps heat up the PET preforms.

6.3 Types of Heat Transfer

There are three ways of conveying heat and all three are present in a reheat stretch blow machine:

- conduction
- convection
- radiation

Conduction occurs when two parts touch each other and heat flows from the warmer to the colder part. An example of this is a warm mandrel transferring heat into the preform neck. Conduction is not suitable for use as a means of reheating because PET like most plastics is a poor conductor and it would take too long to reheat preforms this way.

Convection is heat transfer with participation of air. Heating up of air in the ovens of the blow machine takes place and depending on the temperature of this air it can either heat or cool the outside of the preforms. Air heating is difficult to control, highly dependent on environmental conditions and furthermore heats the outside of the preform more than the inside; just the opposite of what is required.

The output of the oven lamps radiates to the preforms in the form of waves. These waves can be readily absorbed by the PET or penetrate with little absorption. When surface absorption takes place, a large amount of heat is transferred directly to the outside surface of the preform. From there it would then travel by conduction to the inside. This behavior is undesirable since the outside wall would overheat with the inside wall staying colder. Therefore a high degree of surface absorption is not suited for optimal reheating.

6.4 Light Absorption Characteristics of PET

Oven lamps can be adjusted in a voltage range of 0–220 V. On some machines a percentage setting of 0–99% indicates this. One less understood feature of this control is that voltage output affects both temperature *and* wavelength. We measure wavelength for this part of the electromagnetic spectrum in either micrometers (one millionth of a meter) or nanometers (one billionth of a meter) and 1000 nm make up 1 μm. While these lamps all work in the infrared spectrum, this spectrum ranges from 0.7 to 100 μm and it depends on the voltage setting, which wavelength the lamp is emitting. It should be noted that while the lamps work mostly in the infrared spectrum, they also emit wavelengths on either side of this range. Shorter waves of 400–700 nm are in the visible spectrum (Fig. 6.3).

Fig. 6.4 shows how PET absorbs the output of the infrared lamps depending on the wavelength of the emitted radiation. It was recorded by emitting the different wavelengths shown as the horizontal axis through a strip of PET and measuring the wavelength after it bounced off a reflector. To the left of the horizontal axis are the short waves. Here PET absorbs up to 50% of the emitted radiation. Between about 1000 and 2200 nm absorption is about 20% with 80% of the heat being lost. At longer waves to the right of the axis absorption increases again.

High absorption rates would lead to an overheating of the outside skin of the preform, which is detrimental to an optimum heat profile. Therefore the majority of the lamp output should be between 1000 and 2200 nm. This is the output the lamp emits at a voltage of 220–110 V or 100–50%. While this is admittedly a wasteful process, it allows the heat waves to penetrate the preform walls evenly, heating inside and outside walls to roughly the same degree (Figs. 6.5–6.7).

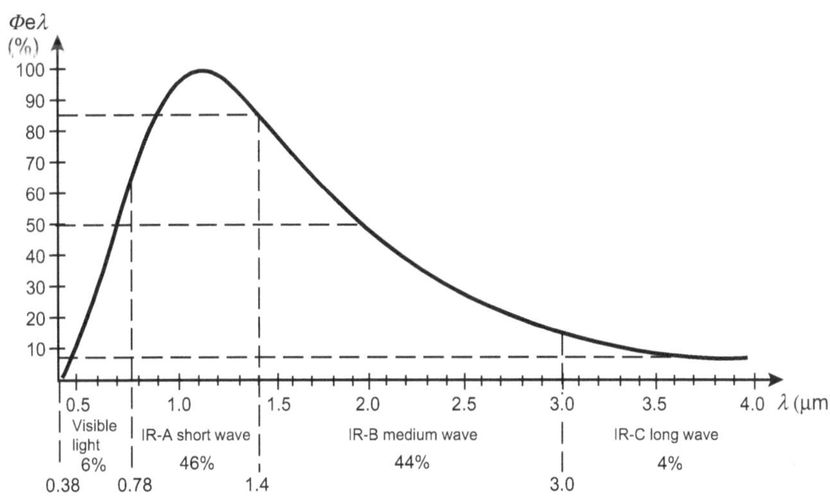

Figure 6.3 Typical range of emitted wavelengths of infrared lamps. Diagram courtesy of Philips.

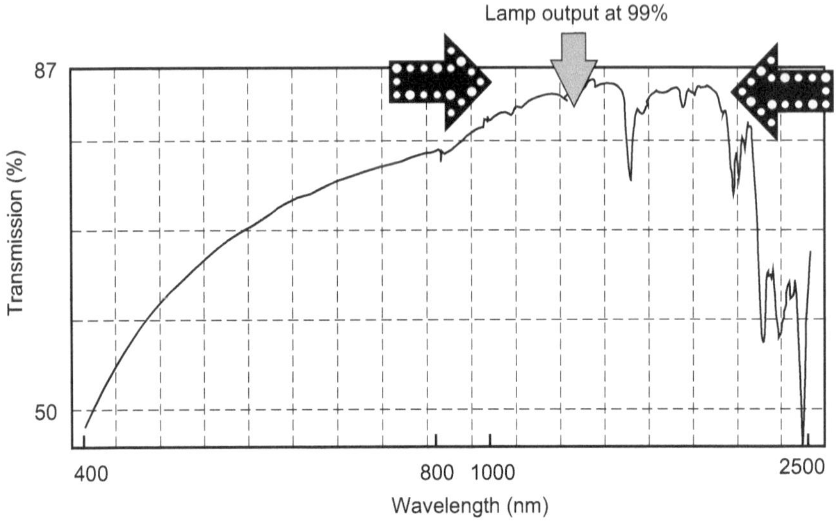

Figure 6.4 Lamp settings change both heat flow and wavelength. Optimal wavelength is indicated between the two *black arrows*.

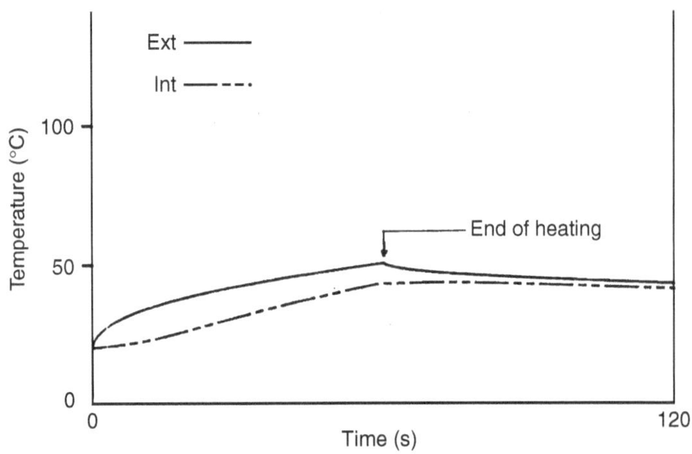

Figure 6.5 Summary graph of empirical temperature data of preform inside and outside wall at 40% lamp output. Inside temperature lags. Graph by Mr. Bonnebat.

Now that the preforms are reheating evenly, the next step is to cool down the outside wall of the preform. This is accomplished by blowing air into the ovens at a rate high enough to lower the temperature to below blowing temperature. This air will absorb heat more from the outside of the preform wall than the inside since it comes more into contact with it. An oven temperature of 85°C (185°F) has been proven to be sufficiently

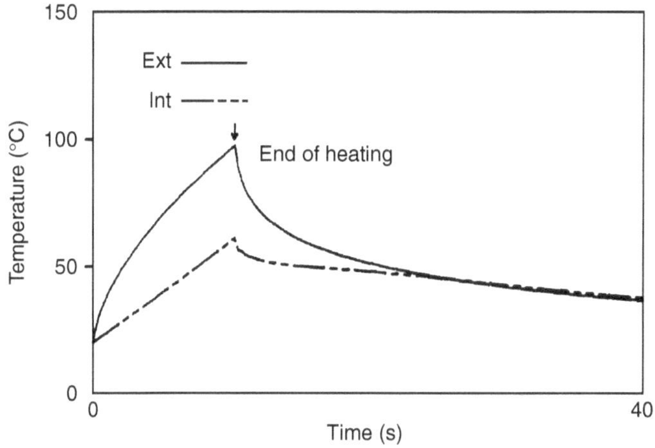

Figure 6.6 Summary graph of empirical temperature data of preform inside and outside wall at 70% lamp output. High absorption at outside wall leads to great difference. Graph by Mr. Bonnebat.

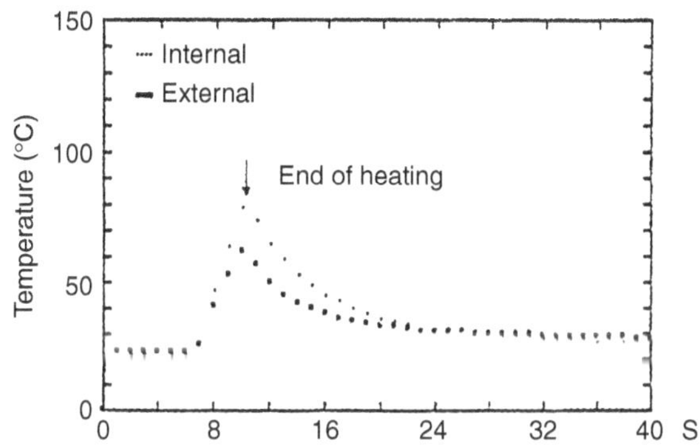

Figure 6.7 Actual temperature data of preform inside and outside wall at 100% lamp output in one position. Inside temperature increases faster due to ideal radiation wavelength and convective cooling of the outside. Graph by Mr. Bonnebat.

low to facilitate this effect. However, oven temperature readings are highly dependent on where the thermocouple is mounted and how many lamps are used. If the lamps closest to the thermocouple are at a high percentage, the reading will be higher just because of this and not because the temperature is actually higher. If the thermocouple was mounted some distance away from the lamps, let us say, somewhere in the exhaust duct it would give a more reliable reading of the temperature we are interested in when it comes to preform heating but would then not give us a good picture of the heat inside the oven. In short, this reading should be regarded with caution.

There are two features a blow machine must have so that air cooling can be effective:

- a thermocouple in one oven measuring the oven temperature and
- a variable speed motor for the air exhaust fan.

The thermocouple must be protected from lamp radiation and be in the air stream, cooling the preforms, not in the one cooling the lamps. The fan should effectively blow warm air out of the system so the oven temperature can be controlled under a variety of process conditions. The reading of this thermocouple must be taken with caution. Depending on how high in the oven it is mounted and whether the top lamps are on or not, readings can differ by a significant amount. The reading is useful to check for differences once a suitable oven temperature has been established but it is not possible to give a recommendation that fits all circumstances.

Equilibration time is another way of evening out differences in temperature through the preform wall. Equilibration happens when preforms leave the oven section on the way to the blow molding station as well as between ovens. Time spend in equilibration depends on the physical length of track between heating and blowing and the throughput rate. Due to more exposure to the environment outside the ovens, outer preform walls tend to cool down more than inner walls thus contributing to the optimal temperature profile. Some machines have extra space in the heating section to allow ovens be moved for optimal equilibration.

It is the lack of sufficient equilibration time on some small, linear machines that often leads to inferior bottle quality.

6.5 Optimal Preform Temperature

Now that we have established the optimal temperature profile within the preform wall, it is time to consider the overall temperature of the preform. PET has a temperature process window from 90 to 115°C.

Higher temperatures cause the material to crystallize. This becomes first visible as cloudiness or haze when the amount of crystallization reaches about 4%. Since crystalline material reduces many bottle properties, haze always causes bottles to be rejected and must therefore be avoided. As mentioned earlier crystalline haze may already be present in the preform and operators need to check their supply of preforms first when this defect occurs.

Temperatures below 85°C (185°F) cause microcracks in the PET structure. These cracks show up as whitish rings, usually in areas with the highest

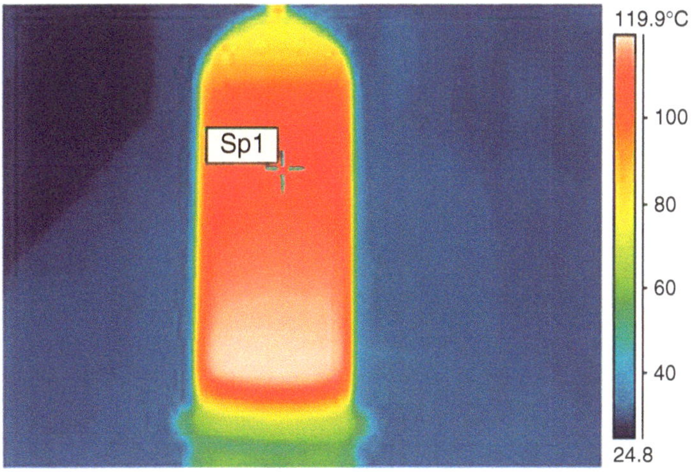

Figure 6.8 Thermo image of preform at the oven exit. Lower blowing temperatures yield better performing bottles. Photo courtesy of Husky Injection Molding Systems.

stretch ratios and also cause bottle rejection. This type of failure is called pearlescense because of the pearl-like appearance of the defect (Fig. 6.8).

Making bottles that show neither crystalline haze nor pearlescense still leaves a process window to work with. At very high stretch ratios there is often only one temperature the preform can be successfully blown at but most other times operators can blow at a certain temperature range. Bottles produced in this range might all look alike but when it comes to bottle properties such as carbonation retention, the bottles blown at the lowest possible temperature before pearlescense sets in are the ones that will perform best. This is because—as shown in Chapter 3, Section 3.2—by lowering the preform temperature the maximum strain within the natural stretch ratio is reduced and the material is forced to orient more. It should now also be apparent that the optimal temperature profile within the preform wall (higher inside than outside temperature) leads to lower overall temperatures since the onset of pearlescense can be delayed: The further stretching inner parts do not fracture as early at the higher temperature.

I should mention at this point that most bottles blown in this matter, that is, at the lowest possible temperature, require a minimum blow air pressure of 35 bar with bar as a maximum. Compressed air takes up approximately 50% of bottle production energy cost (without the cost of preforms) and higher air pressure increases overall bottle cost. Sometimes a compromise between optimal bottle quality and economic considerations must be found. Recycling high-pressure air for preblow pressure is available on some machines and can reduce these costs.

7 The Blowing Process

Chapter Outline

In this chapter, we will be examining what actually happens inside the blow mold during the blowing of bottles. This description is based on materials science, practical experience gathered by interrupting the blow process at various stages, and blow mold simulation software.

7.1 Reheating Preforms

Throughput rate, air flow, and lamp settings all determine the temperature of the preforms before blowing. Throughput rate may be dependent on preform wall thickness if the heating capacity of the machine is limited, or by the required blow process time. For example, a 2-L bottle requires a longer venting time than a 500 mL bottle. Although throughput rate can be adjusted while the machine is in production, it is easier to set the machine at a speed derived from experience with similar bottles before starting up. On rotary machines blow wheel movement and mandrel chain speed are coupled and output rate determines process time that the processor can then subdivided into the different process stages within the confinements of the machine cycle. On some linear machine a process time can be selected and the machine adjusts the output to suit.

Air flow is best left at a medium level and adjusted once some bottles have been blown.

The first step in lamp adjustment is to adjust the height of the oven bank to suit the particular neck design. For this purpose a cold preform is placed on a mandrel and the oven heat shield positioned such that it is just above the neck support ring of the preform. Most oven banks are

Stretch Blow Molding. http://dx.doi.org/10.1016/B978-0-323-46177-1.00007-X

Figure 7.1 Modern ovens allow easy access for adjustment and maintenance. Photo courtesy of Krones AG.

mounted on adjustment blocks that can be vertically moved with threaded rods (Fig. 7.1).

The next step is to examine the preform to be blown. Preform wall thickness ranges from 2 to 5 mm (0.080–0.200 in.) for standard bottles up to 5 L and might not be evenly distributed. Preform designers strive to create an even body wall thickness but bottles such as those with a champagne style base (e.g., beer bottles) need extra material close to the base. These preforms have 4–5 mm wall thickness in the lower fifth of the body, requiring extra heating there.

The initial heat profile will always have a high setting for the first, and sometimes the first two lamps. This will depend on where the transition area between the neck of the preform and body ends, and the shoulder angle of the bottle. Steep angles in the bottle shoulder require less material and the preform must be heated more in the area that will form them.

The remaining lamps will be adjusted to suit the preform body wall thickness.

As stated earlier, there are rotary and linear ovens. Rotary and single-line ovens heat up preforms in one continuous motion, whereas linear ovens may feature a U-shaped track system with no oven module in the turnaround section. The latter arrangement leads to two increments of equilibration time as shown later.

The possibility of overheating the outer preform wall is minimized with the addition of the second equilibration period.

7.2 Blowing Bottles

Fig. 7.2 is a cycle diagram of a typical rotary stretch blow molding machine. Black arrows indicate the duration of the machine function. In rotary machines, all times are actual degrees of rotation of the blow wheel rather than time increments. This has the advantage that certain process characteristics can be described without referring to varying cycle times. A disadvantage of cam-controlled movement comes into play when the machine has to be adjusted for throughput. This might be the case when a thicker-than-usual preform wall has to be blown and the existing oven banks are not able to reheat these at the faster cycle time. For optimum performance it becomes necessary here to adjust the cams for mold opening, closing, and so on since the same number of degrees now means longer times and there is no process reason to slow down the mold closing simply because the machine is running slower. Cam adjustment is more time-consuming and so adds to the changeover time of the machine.

There are two important points that require careful mechanical adjustment:

- The point where the stretch rod just touches the bottom of the preform without stretching it, the so-called "0" point.
- The end position of the stretch rod, the "10" point. This position must leave approximately 1–2 mm (0.040–0.080 in.) between stretch rod and base insert for the preform gate depending on the wall thickness in the gate area of the preform.

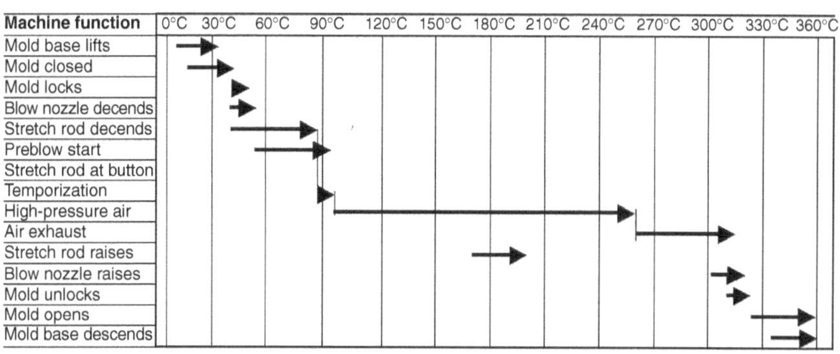

Figure 7.2 Rotary blow molding machine cycle diagram. Rotary machines use increments of one rotation of the blow wheel to control machine functions.

Both adjustments must be made with cold preforms while the machine is operated manually. Exact adjustments are important for proper processing.

For the process it is irrelevant whether the preform is neck up (as in the following drawings) or upside down.

Mold Closed

At the start of the blowing cycle is a temperature-conditioned preform located on a mandrel or placed in the blow mold via grippers (Fig. 7.3).

An O-ring situated at the inside of the neck or at the top thread surface inhibits air exchange between the inside of the preform and the environment. On machines where the preform is taken off the mandrel for blowing, a blow nozzle is pushed into the neck of the preform for this purpose. The mold cooling water temperature will be between 8°C and 65°C (45°F and 150°F) with higher temperatures being more favorable to the process with the caveat that they may increase cycle time.

At 12°C in the cycle the base insert starts lifting. It has to be in position to interlock with the blow mold. To save time, the mold starts closing at

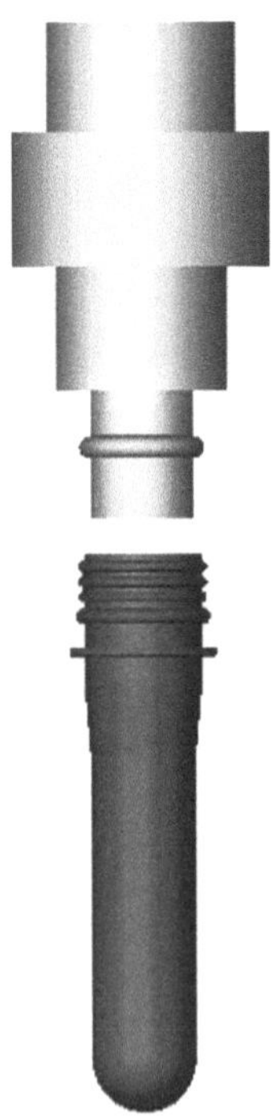

Figure 7.3 An O-ring may seal the preform from the environment. Use and placement of O-rings are particular to each machine manufacturer.

Figure 7.4 Mold assembly at the start of the blowing process.

14 degrees and is closing around the base insert at 39°C. Once the blow mold has closed at 49 degrees and locked at 57 degrees, air pressure inside and outside the preform is equal and the stretch rod tip has moved to a position very close to, but not touching, the inside of the preform base. The locking is often checked with the help of a switch at 69 degrees (Fig. 7.4).

Stretch Rod Engages

Both blow nozzle (if present) and stretch rod start moving at 51°C. The preblow pressure may start as soon as the blow nozzle is fully down but there are some delays before full preblow pressure is actually acting on the preform. For example, the blow valve has a certain delay because mechanical parts in the valve must physically move before air can actually pass through. The air has to cross the distance between the valve and the cavity and the closer the valve is mounted the faster the process can start. Once inside the cavity, the inrushing air will have slightly decreased in pressure and it is a fraction of a second before pressure is built up again. These delays combined, take approximately 0.05–0.15 s, mostly dependent on

Figure 7.5 The stretch rod engages the lower part of the preform and may pull it out before stretching other parts.

the distance between blow valve and cavity. Stretch rod action may be delayed to allow pressure to enter at the same time as the stretch rod moves. Depending on the preform temperature the stretch rod may stretch the gate section of the preform first because of mechanical stress acting on this surface before reaching other parts of the preform (Fig. 7.5).

Preblow Engages

As the stretch rod travels further preblow pressure reaches its maximum value of 5–20 bar (70–290 psi). The rubber-like state of the polyethylene terephthalate (PET) will now be inflated into a bubble at the weakest point. This is usually one third to halfway down the heated length of the preform. With properly designed preforms, this yielding phase (Chapter 3, Section 3.2) will come to an end before the bubble reaches the mold walls, thus allowing room for the material to orient more later. As the material extends and reaches the end of its natural stretch ratio the ensuing molecular orientation increases the mechanical strength of the stretched areas. Now areas bordering on the already stretched part are weaker and will therefore

Figure 7.6 Preblow pressure creates a bubble not unlike the way a balloon inflates.

start to extend. If the preform is at the same temperature throughout the walls, the bubble will develop from the top down (Fig. 7.6).

The stretch speed is an important aspect of proper bottle blowing. Too low a speed can lead to preforms cooling down to a temperature approaching the glass transition temperature where they will not stretch evenly. Higher speeds also improve molecular orientation.

Stretch Rod at Base Insert

As the stretch rod races toward the base insert of the mold, more and more sections of the preform start yielding, thereby enlarging the bubble. The speed of development and the extent of the bubble depend on the material's intrinsic viscosity (IV) and temperature. The situation might be as depicted as to the left (Fig. 7.7).

The stretch rod reaches the end position at approximately 90 degrees in the cycle depending on the length of the bottle. It is advantageous in some cases to hold the preblow pressure for another 0.05–0.1 s allowing the bubble to extend further down and so pushing more material to the outer rims of the bottom of the bottle. For many other bottles it is best to trigger

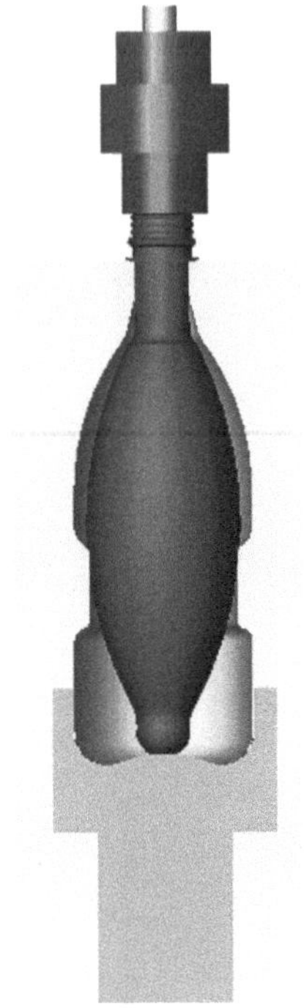

Figure 7.7 When the stretch rod has fully extended the bubble has grown to fill most of the cavity.

the high pressure as soon as the stretch rod is fully extended. On some machines there is a special timer called "temporization" which delays the onset of high-pressure blow. Other machines feature simply an adjustable switch indicating the stretch being fully extended, and otherwise rely on the slight delay of the start of blow for the reasons described earlier.

The preform is now ready for orientation (Fig. 7.8).

High-Pressure Blow

Blowing may start as early as 95 degrees or as late as 120 degrees depending on the air delay or the amount of temporization. The impact of air at 40 bar (580 psi) pressure is dramatic: it takes only around 0.02 s to change the bubble into a fully formed bottle. Air pressure forces the material against the cooler walls of the mold cavity. The material must be cooled below the glass transition temperature and another benefit of running cooler preforms is that the required cooling is minimized, allowing faster cycle times. It should be noted that the stretch rod has to hold the preform gate firmly in the center of the base insert. Any deviation will result in uneven bottle wall thickness. Some linear machines use the air pressure to the stretch rod cylinder as speed control.

Figure 7.8 The bubble can be extended further by slightly delaying the onset of high-pressure blow.

Reducing this pressure to too low a value might allow movement of the preform bubble under the impact of high-pressure blow air and move the gate!

Some machines offer a system to circulate high-pressure air through the blow mold further reducing cycle time. In either case blowing finishes around 300 degrees (Fig. 7.9).

Mold Opening

The stretch rod moves out of the blow mold starting as early as 120 degrees Venting is initiated with the end of blow. To facilitate this aspect, a venting valve allows air to escape via a large diameter hose or pipe ending in a suitable silencer to keep noise pollution in the plant down to a minimum. Venting finishes at 342 degrees and only then can the base be lowered. Lowering the base earlier would lead to the deformation of the bottle base known as rocker bottom. However, during venting, the clamp unlocks and at 320 degrees starts to open. This is possible because at that time pressure in the blow mold has significantly dropped. Remaining pressure may slightly bulge the bottle side walls outward but this will not lead to a permanent deformation. At 360 degrees the mold is open, the base insert is down and a new cycle begins (Fig. 7.10).

Figure 7.9 High-pressure blow moves the preform the rest of the way to the walls of the blow mold.

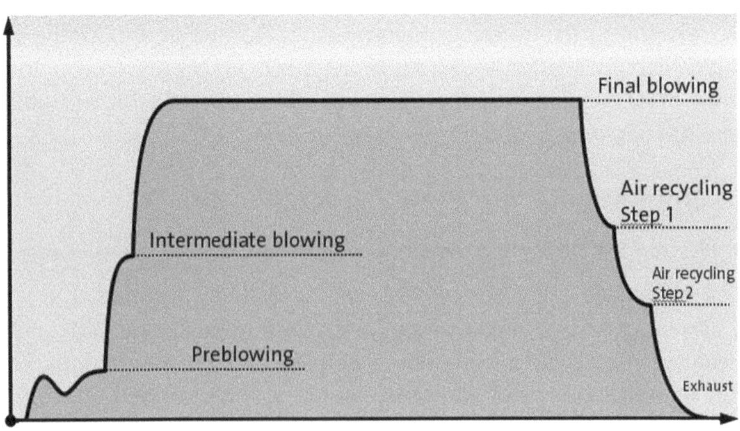

Figure 7.10 Pressure curve inside the blow mold here shown with optional air recovery. Diagram courtesy of Krones AG.

7.3 Air Valve Control

At this point it may be beneficial to understand the functionality of the air valves that are so crucial to the proper timing of the blowing process. Today, most air valves used in blow molding come in blocks of three or four: preblow, blow, exhaust, and a second exhaust valve or one used for air recycling. These valves are electrically triggered pneumatic slave valves as shown in Fig. 7.11.

The valve on the left is electrically operated. When it switches, it opens the path of the air through the valve on the right.

These valve blocks are mounted as close to the cavities as mechanically possible to minimize "dead air" loss. Modern valves contain pistons made from PET or other plastic that is durable and light to allow switching in milliseconds. This is one of the "secrets" of the faster machines!

Here are the different stages of valve engagement throughout the cycle. P1 stands for preblow, P2 for high-pressure blow, and EX for exhaust (Figs. 7.12–7.18).

When troubleshooting the air pressure circuit, it is best to check the pressure diagram of the cavity that produces defective bottles. All modern rotary machines have these. A "sticky" preblow valve is often the cause of bottle problems and these screens show this very well.

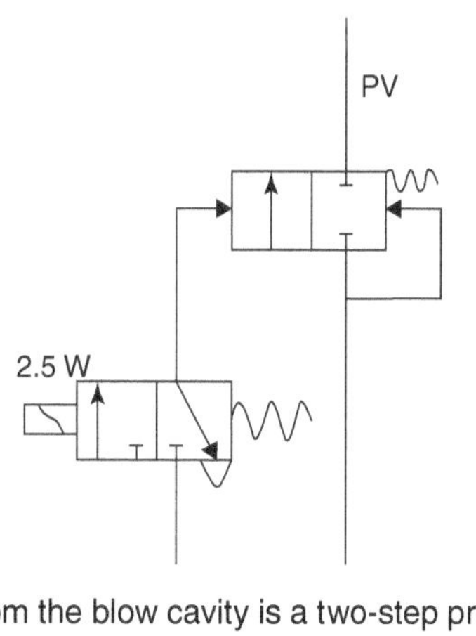

Figure 7.11 Air to and from the blow cavity is a two-step process whereby an electrically operated valve switches an air-operated one.

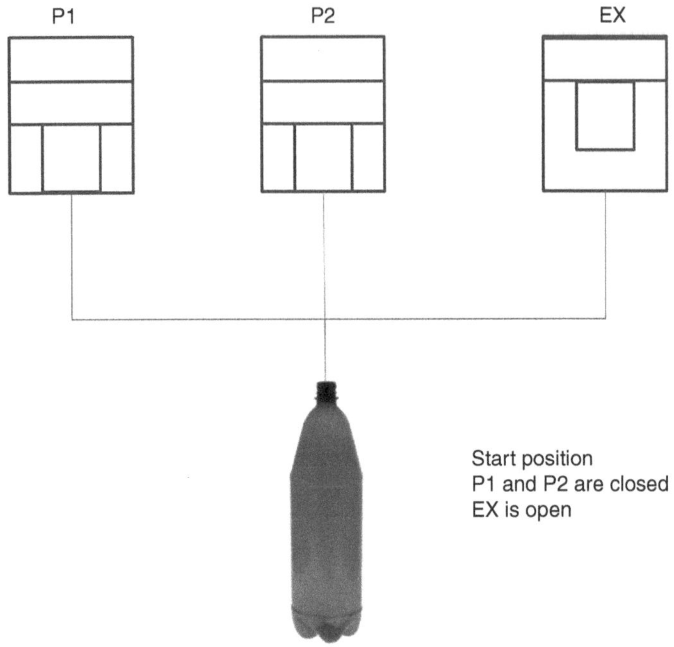

Figure 7.12 Air step 1. At the beginning of the cycle P1 and P2 are closed and EX is still open from the previous exhaust cycle.

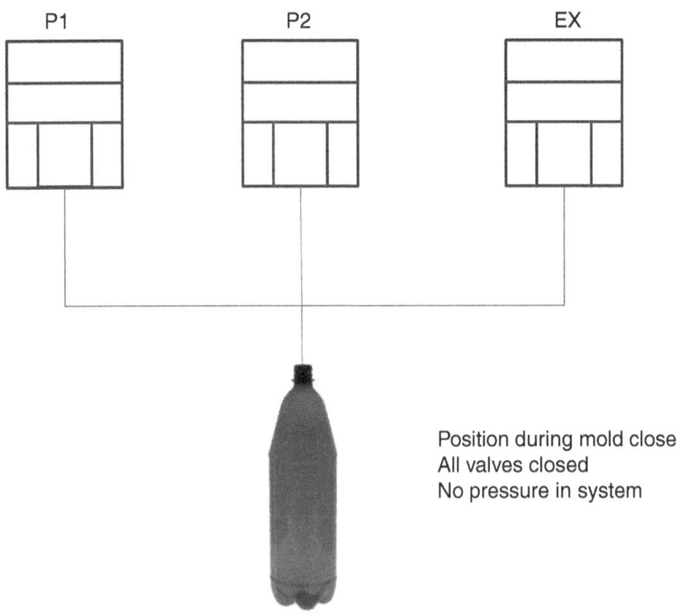

Figure 7.13 Air step 2. The first event is the closing of the exhaust valve.

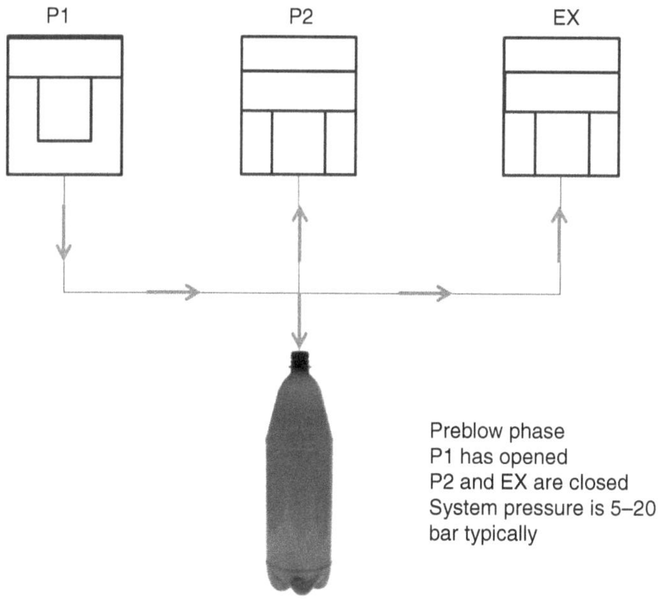

Figure 7.14 Air step 3. Now the P1 opens and the preblow phase begins.

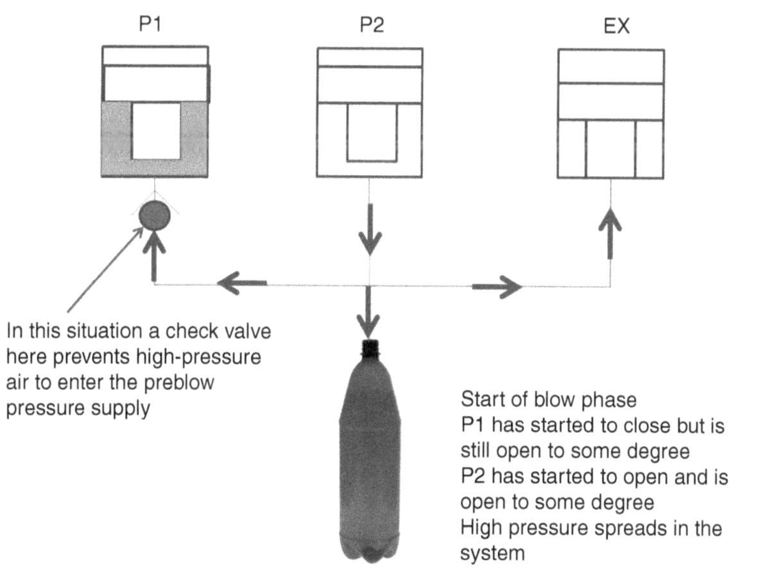

Figure 7.15 Air step 4. To initiate high-pressure blow, P2 opens at the same time as P1 closes. To prevent high-pressure air from getting into the preblow circuit, a check valve is mounted into the preblow line.

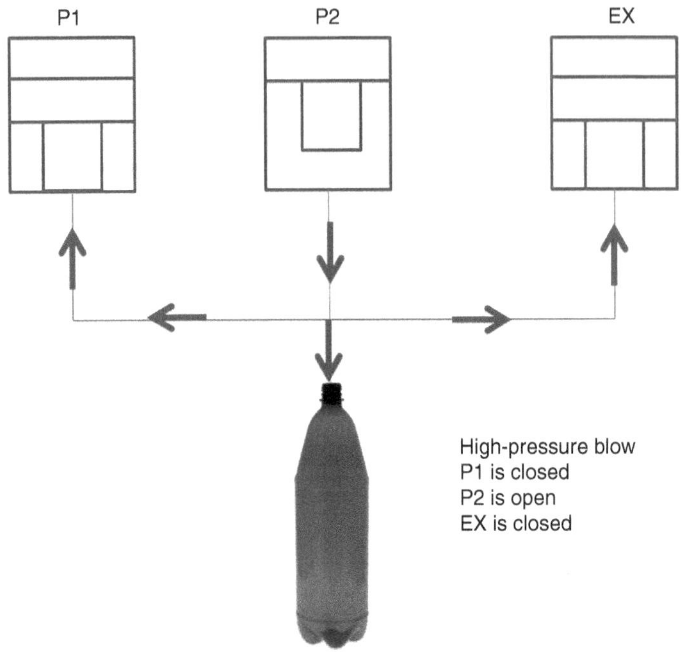

Figure 7.16 Air step 5. The high-pressure valve is fully open and the container is blown to its finish dimension.

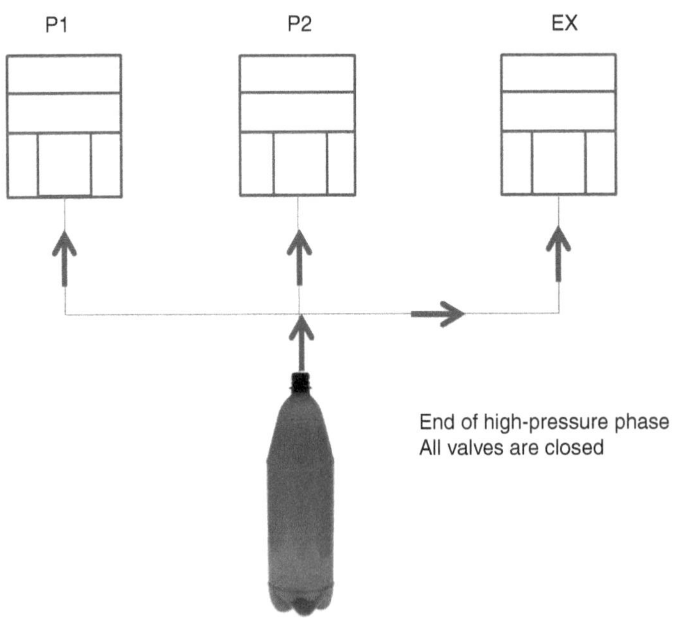

Figure 7.17 Air step 6. Before the exhaust valve can open, the high-pressure valve must close first.

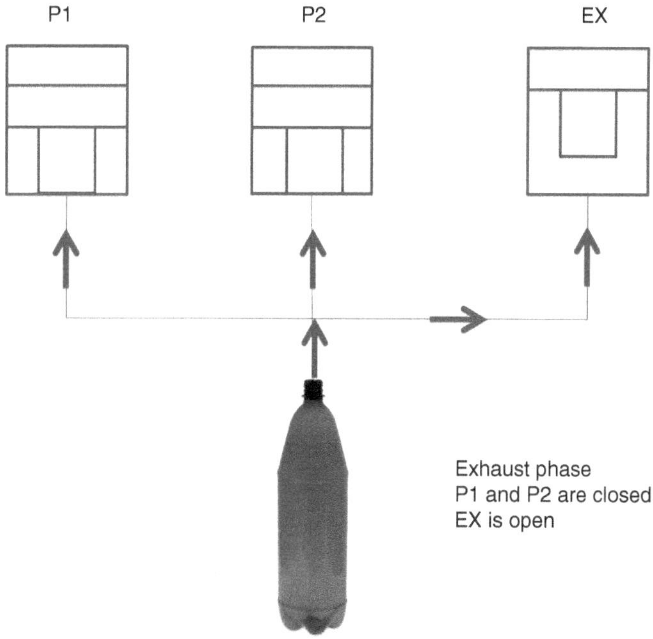

Figure 7.18 Air step 7. Now the exhaust valve opens and all air escapes.

8 Injection Stretch Blow Molding Machines

Chapter Outline

As the name implies this process involves injection of preforms as well as blowing of bottles in the same machine. Preforms are not cooled down completely after injection; instead residual heat inside the preform allows blowing without reheating. Obviously, this makes it inherently more complex than the two-stage process. Processors must know both parts of the process and, as we will see, there is also some interaction present between injection and blow adding to the overall complexity.

Machines of this type are commonly (but by no means exclusively) used in the cosmetics, personal care, and household markets. They are hardly used in the water and carbonated soft drinks (CSD) market as their outputs are limited. There are actually two different types of machines that can be summarized under injection stretch blow molding (ISBM):

Stretch Blow Molding. http://dx.doi.org/10.1016/B978-0-323-46177-1.00008-1

Figure 8.1 A wide variety of bottles can be made in this process. Picture courtesy of Nissei ASB Company.

- Indexing with three or four stations where the preform stays in the neck insert, going through injection (conditioning with four-station machines), blow, and ejection.
- Various designs where the preform is completely removed from the injection tool, put on mandrels, heat-conditioned, and then blown. This process may be called integrated two-stretch blow molding (ITSBM) (Fig. 8.1).

8.1 Four-Station Machines

Preforms and bottles are oriented vertically and in rows, typically one row only but there are also machines with two rows of up to 16 cavities making 32 cavities the current maximum number. They are spaced not with respect to the preform dimension but to the maximum bottle dimension as the pitch between the cavities does not change within the machine. A 90-mm wide or round bottle requires spacing of about 100 mm in both preform and blow tooling to allow some metal between the blow cavities. A particular machine can produce a 300-mL bottle in 16 cavities or a 15-L bottle in single cavity with various bottle sizes and respective cavitation in between (Fig. 8.2).

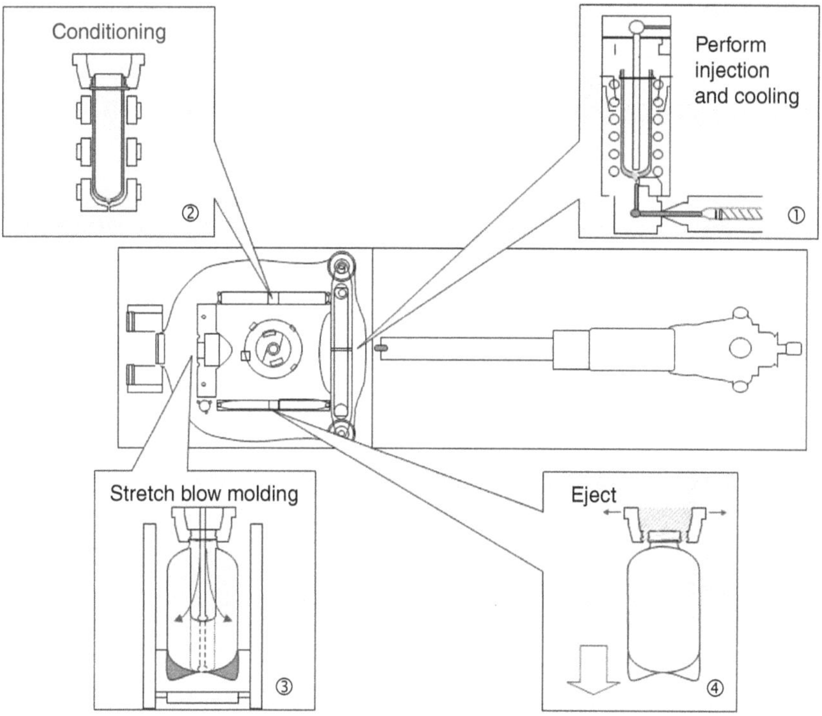

Figure 8.2 Four-station ISBM process. Diagram courtesy of Nissei ASB Company.

In order to engage the injection tooling vertically, the following two movements are required:

- Either the cavities move down and the cores move up with the neck inserts staying in place before rotating. This requires the cavity block with the hot runner underneath it to disengage from the extruder that is moved back. One issue with this arrangement is that air or pollutants can enter the melt stream when extruder and hot runner are not connected.
- Or both cavities and neck inserts move up leaving the cavities in place. This is a superior solution as the aforementioned problem is not an issue (Fig. 8.3).

A machine with four stations rotates 90 degrees moving preforms and bottles around every cycle, whereas a three-station machine rotates 120 degrees. Orientation of each station is vertical allowing a very compact

Figure 8.3 Injection tools are oriented vertically with cores on top and hot runner and cavities on the bottom. Picture courtesy of Nissei ASB Company.

footprint. Other advantages of this system besides the ability to make both preforms and bottles in one machine are as follows:

- Flexibility to make bottles of different shapes with good wall distribution.
- Production of blemish-free bottles. Since preforms are not touched, bottles have no marks when they leave the machine. Important feature for cosmetic bottles.
- Zero contamination between preform and bottle stage. This may decrease sterilization measures in the filling plant.
- A neck support ring is not necessary, also interesting for cosmetic bottles.
- Thermal efficiency; preforms are cooled down only to blowing temperature of about 100°C (212°F), saving energy needed to cool them to about 60°C (140°F) necessary for two-stage molding.
- Automatic neck orientation; some caps require the neck to start in a particular position with respect to an oblong bottle shape. This requires an expensive device in two-stage molding but is "free" in ISBM as the neck start can be chosen deliberately and the preforms are held in place between injection and blow without spinning.

- Vertical injection tooling leads to longer tool life because gravity does not try to pull tools off-center as is the case with horizontal injection machines.
- Many custom bottles have special neck finishes for which it is hard to find preforms on the open market. In this case, a single-stage solution may be the most cost-effective solution as it is often cost-prohibitive to build and run injection tooling for preforms of low volumes.
- The inside of the preform is always hotter than the outside. This is of great advantage as the inside always has to stretch more than the outside (Chapter 6, Section 6.2). The reason why this is happing in single stage is that the cooling area on the core side is always smaller than on the cavity side and often with less flow (see Chapter 9, Section 9.6 for a more detailed explanation). This explains the ease of forming difficult container shapes.

There are also a number of disadvantages mentioned as follows:

- Cycle times are relatively slow even when comparing the same thickness and weight preform molded on a modern injection machine. Part of this has to do with the way the machine rotates requiring several tool movements that accumulate dead time.
- Long changeovers: Injection tooling is very cumbersome to remove as each injection core must be replaced individually. Machines are also difficult to access making it more time-consuming to work on them.
- Each blow cavity requires one injection cavity. Because injection usually takes 2 to 3 times longer than blowing the blow section is idle for half to two thirds of the cycle time. In two-stage molding it is much easier to match the output of an injection machine with that of a blow machine. For example, a 72-cavity injection machine running a 8-s cycle time produces around 32,000 preforms/h. A blow machine with 16 cavities running at 2000 bottles/cavity per h will be the ideal candidate to match. Both machines can run at full capacity, a great advantage given capital and footprint limitations.
- As explained in greater detail in Chapter 9, Section 9.7 processors are not in complete control of the preform temperature profile limiting the uniformity of bottle wall thickness and reducing bottle quality (Fig. 8.4).

Figure 8.4 Blow station (left) and ejection station (right). Picture courtesy of Nissei ASB Company.

8.2 Machine Controls

Melting in Extruder

The resin that enters the extruder throat is a mix of crystals and amorphous parts. In order to melt the resin the extruder must

- heat and soften the amorphous fraction and
- melt the crystalline fraction.

By rubbing the pellets against each other and against barrel and screw the extruder generates the necessary shear heat for melting. All crystals must be melted or they will become nuclei (starting points) for crystallization in the preform.

Heat transfer from barrel through heater bands is only about 30%,may even be negative in some zones. Negative heat transfer would be the case when the temperature readout of an extruder zone is higher than the set point. In this case the friction inside the barrel is so high that it actually overheats the barrel and must be cooled down to maintain the temperature that is selected. This usually happens at the end of the barrel in the so-called metering zone of the screw. Most heat (about 70%) comes from pellet inlet temperature (dryer) and from friction (screw and barrel). The operator has control over the heats, the screw rotational speed, and the back pressure during screw rotation, which is called recovery.

While temperature screens differ from machine to machine, they all convey the same information.

They may show:

- the location of the heater band (usually going right to left)
- °C or °F
- the set point
- the actual temperature
- a display of a temperature without set point is the temperature of the incoming resin as measured just above the extruder throat. A temperature of about 165°C is optimal for polyethylene terephthalate (PET) processing
- the percentage of power the controller puts out to the heater band. For example, if this value is 40%, the heater band is on for 4 s, then off for 6 s. The controller will use a value that is best suited to keep the heater band at the set point. This is regulated by a so-called PID loop and all controllers use some form of this control program.

A typical temperature profile starts at 270°C (518°F) at the feed zone and increases to 285°C (545°F) toward the extruder nozzle. This can be used for most PET applications. The extruder cannot be started until all heaters are at the set point and a soak timer has timed out. The machine heats have to be enabled either by a physical switch or by a switch on a screen before the machine starts heating. If a "soft start" feature is available and is selected heater bands heat up slowly.

Soak time may be available on some machines and is the time between the moment when the last heater band has reached its set point and the moment the machine allows the extruder to start. Soak time is different for extruder and hot runner. A longer soak time does not harm the process but too short of a soak time may. Use 30 min for the extruder and 15 min for the hot runner. If a "standby" function is available it allows the dialing in of a second, lower set point. When the machine is expected to be down for longer than 30 min, this feature is used to prevent material from burning without turning heats all together off.

Older machines may not have a protection against a "cold start," which is the (often accidental) turning on of the extruder before the heats are up and have had time to soak. This will usually break the screw at the thinnest point in the feed section. On these machines a note should be kept on the machine during the heating up process to indicate when it will be safe to start the extruder.

8.3 Injection Controls

All machines on the market use reciprocating screws. That is, the screw turns and pushes material to the front of it while retracting backward in the barrel. This is called recovery. During injection the screw moves forward pushing material into the hot runner while a check valve located at the front of the screw prevents material from moving back.

The parameters that can be adjusted on the machine are discussed as follows:

- injection pressure
- injection speed
- transition point or switchover point
- hold pressure
- hold time
- cooling time
- cushion length

Injection Pressure and Speed

Processors can dial in injection pressure either from the screen or manually on a pressure relief valve. A maximum of 100 bar (1500 psi) is recommended for PET to avoid shearing the material too much. This is hydraulic, not material pressure and the difference should be explained. The hydraulic injection piston has a 5–7 times bigger area than the screw area leading to pressure intensification. For example, a 200-mm injection piston has an effective area of 31.4 cm^2. The 80-mm screw is connected to have an area of 5.024 cm^2. Using 100 bar in the piston results in a force of 3140 kg acting now on the smaller screw area. This leads to the screw experiencing a pressure of 3140 bar/5.024 cm^2 or 625 bar. This is the pressure the material is subjected to and 700 bar (10,000 psi) is the recommended maximum for PET.

It is possible to increase this pressure slightly during start-up to get the process going but many machines run well below the maximum. A notable exception is very thin preforms (<2.3 mm), as these require fast injection and with it high pressure to prevent freezing off in the cold tool. Pressure is also not actually regulated by the pressure setting. Instead, this setting is just limiting the pressure to the dialled-in value, draining oil to tank when the actual pressure reaches the selected threshold. What actually creates the pressure is the speed of injection. The faster the speed the higher is the resulting pressure. Speed should be chosen so that the maximum pressure is not reached. Most machines show the

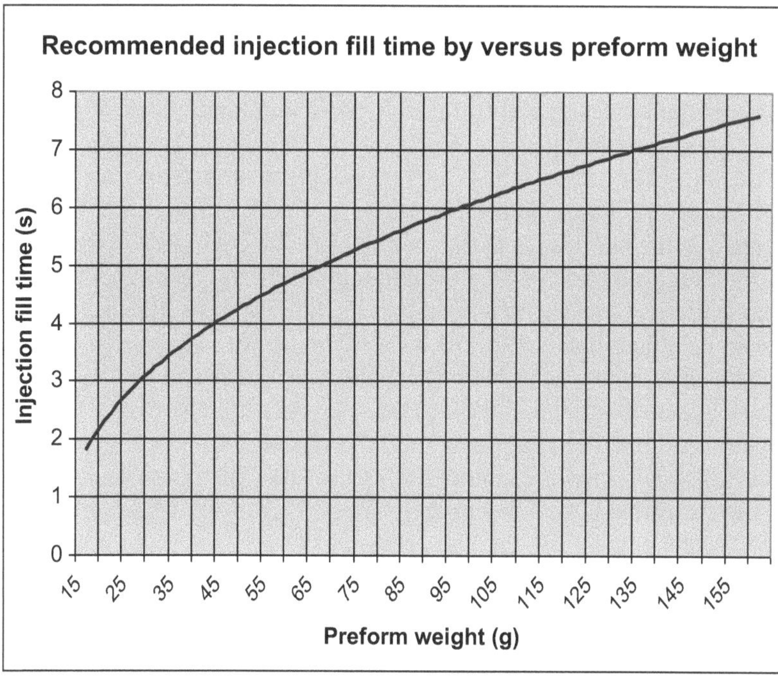

Figure 8.5 Recommended injection fill time versus preform weight. This graph should only be taken as a general guideline. Other process considerations may force processors to change the time significantly.

actual pressure on the screen while on some it can be read on a dial gauge at the extruder.

Injection time is the time between start of injection and when the screw reaches the transition or switchover point. The actual injection time is not a time that can be dialled in. Instead, it is the result of the speed setting and how long it takes to move the material from start of injection to the transition point. This in turn depends on the chosen speed setting and melt viscosity. As discussed earlier, melt viscosity is a function of the material intrinsic viscosity (IV) and melt temperature. Fig. 8.5 gives a guideline what injection time should be for PET.

In order to get to these times there are usually three or more speed settings that the processor can use. These are mostly open-loop controls that can be set depending on the screw position. On the machine a proportional valve receives speed percentages as variations of milliamps and allows a proportional amount of oil flow into the injection device. Recommended is a slow start, a fast middle, and then a slowdown before the screw reaches the transition point.

Choosing the Transition Point

As explained earlier the transition point should be chosen as that point when the injection cavity is completely filled with material. There are several ways to calculate it.

The first takes advantage of our knowledge of melt density versus solid density. For PET these values are 1.16 and 1.335 g/cm^3, respectively. We assume that the material is at the melt density during injection and at solid density after hold and cooling, even though this may not be that correct as explained later. Because melt density is about 87% of solid density our initial shot calculation should allow 87% of the stroke to happen in injection and 13% of it during hold. The 13% hold phase stroke allows the material to grow in density, that is, shrink to the solid density. The reason this is not as accurate is that the material, especially the melt front, cools down during injection and increases in density as a result. This is especially true for thin preforms that cool down quickly. The 13% density difference may be as low as 8% or even 6% thereby changing the necessary settings.

Another way of finding the transition point is to observe hydraulic injection pressure. During injection the cavity is empty (except for air) and there is little resistance to the flow of plastic into it. Once the cavity is filled with material injection pressure will increase immediately as the material now blocks further injection. An experienced processor can thus determine at which point the cavity is filled.

A third way is to calculate the screw stroke with the screw diameter, the melt viscosity and the shot weight as parameters.

Hold Pressure and Time

During hold time the material shrinks to solid density as explained and the main purpose is to allow some material to enter the cavity to compensate for this shrinkage. Most machines have three hold pressure and time settings. The first is responsible for the neck area of the preform, the second for the body, and the third for the gate. A general guideline would be to set the three timers to equal time and the pressures to 60, 50, and 40%, respectively of the actual injection pressure. Fig. 8.6 gives a guideline for the total hold time as a function of preform wall thickness.

Cooling Time

After the hold timers have timed out cooling time starts and the screw moves back relieving pressure in the hot runner and cavities. Cooling time should be chosen to allow the preform to shrink away from the cavity for easier preform removal and to control gate crystallinity. Fig. 8.7 shows recommended times depending on wall thickness.

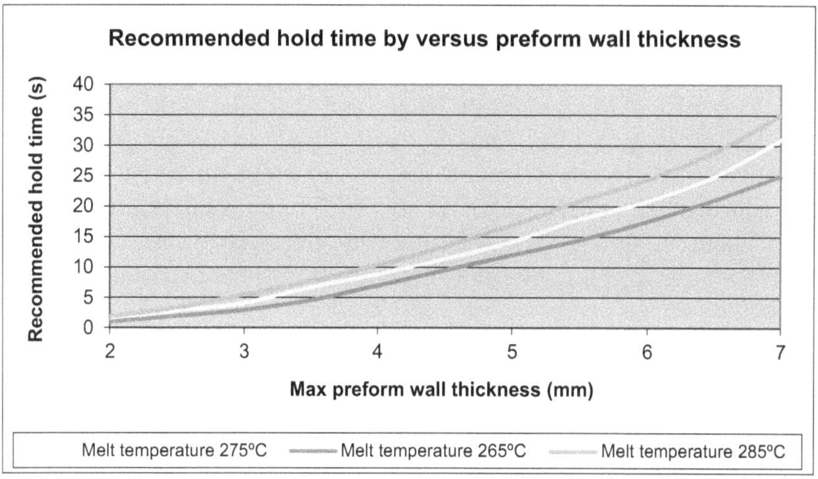

Figure 8.6 Recommended hold time versus preform wall thickness.

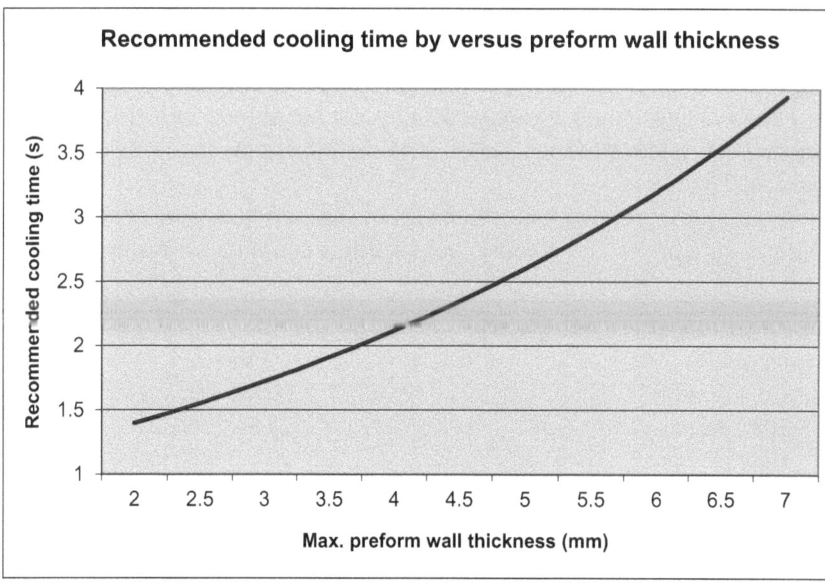

Figure 8.7 Recommended cooling time versus preform wall thickness.

Cushion Control

The screw should not bottom out at the end of the barrel with each stroke; instead there should be what is called a cushion, that is, some material is left to cushion the screw stroke. Typically 5–10 mm is chosen but in single-stage processing a larger cushion may be selected to give the material more time to homogenize.

A Practical Example

This job is for a 60-mm screw machine, running a eight-cavity tool with a preform/bottle weight of 23 g. Total shot size is therefore 184 g. At a melt density of 1.16 g/cm^3 the screw stroke is 184 g/1.16 g/cm^3 per ($6^2 \times 3.14/4$) = 5.6 cm or 56 mm. This is the stroke the screw should traverse during injection. Since this is only 87% of the total, the stroke to the end of hold is 56/87 × 100 = 64 mm. In other words, the screw should traverse 8 mm during hold. Choosing a 5-mm cushion the total stroke setting should be:

Shot size (the point where the screw starts):	69
Transition point:	13
Screw stops at:	5

Speed settings could be 30% from 69 to 60, 50% from 60 to 30, and 25% from 30 to 13.

Initial speed settings will result in a certain injection time and need to be adjusted to come close to the above-mentioned recommended injection times.

These values can be a good starting point but should be verified on the machine once it is running. Other considerations may take precedence as explained later.

Assuming a preform wall thickness of 3.2 mm and an observed injection pressure of 50 bar (700 psi) hold settings could be as follows:

Total hold time (from the graph):	5.1 s
Hold times:	1.7 s each
Hold pressure 1	30 bar (435 psi)
Hold pressure 2	25 bar (360 psi)
Hold pressure 3	20 bar (280 psi)

Sink marks in the neck area may require higher hold pressure 1 and/or time settings. Gate appearance will be controlled with hold pressure 3.

8.4 Interaction Between Injection and Blow

It should be obvious at this point that the injection process has an impact on the blow process. Different hold and cooling times change preform temperatures and this in turn will change the way the bottle blows, especially in areas that are close to the upper and lower limit of PET's ability to stretch.

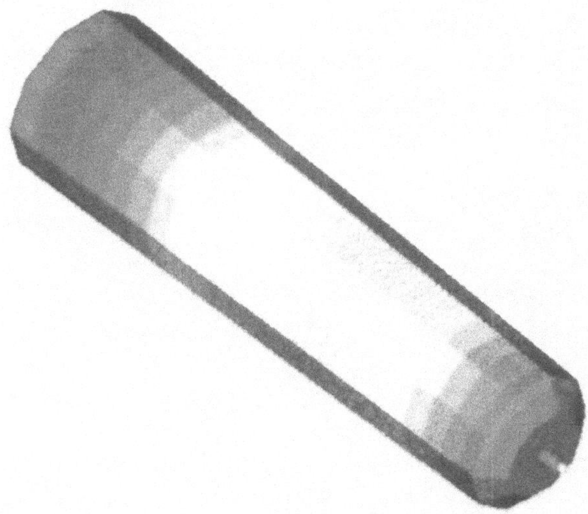

Figure 8.8 Computer simulation of preform temperature.

But that is not all. As previously mentioned, one disadvantage of ISBM is that processors are not in complete control of the temperature profile of the preform. Because the melt in the neck area had a longer time to cool down compared with the melt in the gate area it is also cooler and we may find a temperature profile similar to the one in the graph as shown in Fig. 8.8.

In this simulated graph a temperature gradient from the hot gate to the cooler neck can be clearly shown. Actual temperatures and differences depend on process conditions and are therefore not shown here.

This heat profile will result in a relatively thin bottle base with more material toward the neck because PET will stretch first in the hotter areas. To counteract this behavior preforms are designed with thinner gate areas and/or more cooling and thicker areas toward the neck. While this may sound counterintuitive, a thicker wall in the preform neck area leads to a thinner area in the bottle neck. This is because the thicker preform area retains more heat and will then stretch more resulting in an overall thinning.

For the processor there is another way of changing this temperature profile. It relies on the observation that faster injections will lead to less difference between cooler and warmer areas because the material injected first has less of a time difference spent in the cold mold than with longer injections. Overall, the temperature will be higher but the temperature gradient from hot to cold is less. This will then result in thicker bottle bases.

Of course the opposite is also true, to increase material thickness in the bottle base a slower injection may be selected.

8.5 Conditioning

Four-station machines have the ability to condition preforms after injection and before blow. There are two devices on these machines:

- heater bands that move around the outside of the preforms and heat them without touching them and
- temperature-controlled conditioning cores that move inside the preforms and touch them.

Depending on the length of the preform there can be up to three heater bands that are thermally insulated from each other and that can be adjusted with separate temperature controllers. Heat transfer is limited as the air around the preforms does insulate them to a large degree but they have some effect and are mostly used to heat the preform portion below the neck in order to get material away from the shoulder of the bottle (Fig. 8.9).

Figure 8.9 Conditioning station machine has two rows of tooling to double output. Conditioning is shown on the left with the blow station on the right. Picture courtesy of Nissei ASB Company.

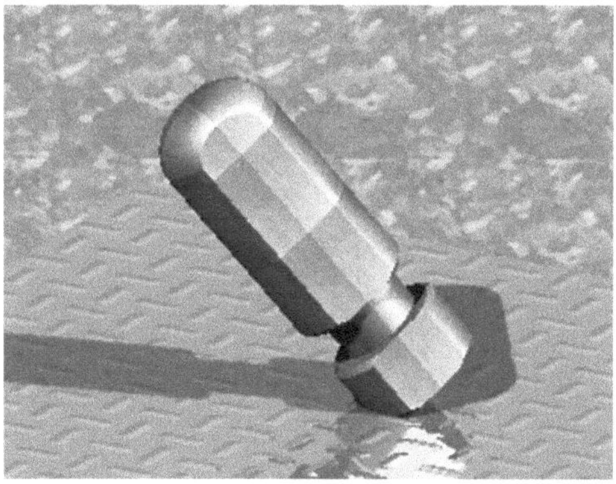

Figure 8.10 Conditioning core for oval container. Conditioning cores can be shaped in a variety of ways to achieve excellent wall thickness distribution in the blown bottle.

The conditioning cores on the other hand are very effective and are the main reason oblong shapes with very good wall distribution can be manufactured (Fig. 8.10).

The function of the conditioning core is to cool down areas that are expected to blow thin compared to areas that will be thick. A typical example is an oval bottle. The short sides of the container would be thick if the preform was heated evenly. Cooling the areas of the preform that will form the far sides of the container down has the effect that these cooler preform parts pull the warmer areas apart thinning them out in the process. The cores are usually run with temperature-controlled water. Thermolators keep the temperature between 55 and 75°C (130–70°F). Conditioning cores are made from aluminum and can be shaped easily into a variety of shapes for oval, rectangular, or other oblong containers. They can also be used to cool down a hot bottom area leaving the top part of the preform untouched. Used in this way there are another way of overcoming the often disadvantageous heat profile that hot runners impart on preforms. While it is not optimal from a process point of view to heat the outside and cool the inside of the preform, the inside temperature of the preform is substantially higher than the outside as explained in Chapter 10, Section 10.6 and therefore has already the optimal profile.

8.6 Container Blowing

Blow molds move horizontally and are similar to molds used in the two-stage process. Since they are idle for around half the time, some companies just cool the back plates they are mounted on, thus saving the drilling of

Figure 8.11 Blow molds, like in this double-row machine, move horizontally. Picture courtesy of Nissei ASB Company.

water lines in the molds. This works because PET is only at about 100°C (212°F) at the time of blowing and the large aluminum body of the molds acts like a giant heat sink and dissipates the preform heat. As cycle time is not an issue they can be closed for a few additional seconds as they warm up over time. It is actually beneficial for the process if the molds warm up to about 60°C (140°F), the material flows easier and blown containers are already stress-relieved to some extent.

Stretch rods enter the preforms from the top controlled by pneumatic cylinders. Most machines use long hoses to carry blow air to the cavities, air that has to be exhausted every cycle. This makes the machines large air consumers and contributes to operational costs. On most machines there are only three timers for all cavities that control stretch rod delay, preblow delay, and blow delay. All three are triggered from the mold close signal (Fig 8.11).

Timers offer a lot of flexibility as they are easy to change from the screen and the operator has full control over the sequence of events. For example, he/she can bring in preblow air before the stretch moves. One disadvantage with this approach is that stretch rods will not always be in the same position when preblow or blow air enter the preform. This is because preforms offer varied resistance against the stretch rod action depending on their temperature, which varies in a typical day/night temperature cycle.

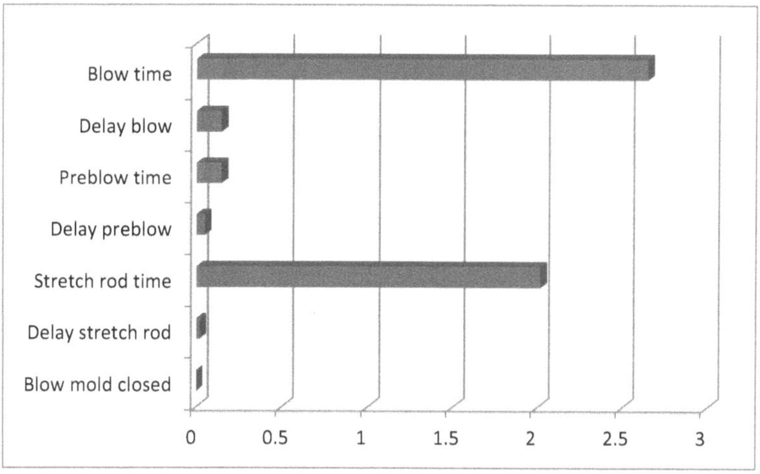

Figure 8.12 Most single-stage machines use timers to trigger the various blow functions.

While these variations are small often the process window on multicavity tools is not very wide and even small changes can throw the process off in at least some of the cavities. Another problem is that the operator does not know when the stretch rod is fully extended. As outlined in Chapter 8, the stretch rod must be fully extended before high-pressure air is introduced. Otherwise, the gates will go off center. The only way of knowing when that point in time happens is to reduce the delay blow air to the point when all gates are off, then add some time to it.

A better albeit slightly more cumbersome way is to time the start of preblow and blow air from switches that are mounted somewhere along the vertical axis the stretch rods have to follow. This assures both that the stretch rod is always in the same position when preblow air comes on and that it is fully extended before blow air enters the cavity. It is therefore the preferable method. Some machines use both switches and timers whereby the switches trigger the timers and this set up offers even more flexibility (Fig. 8.12).

8.7 Hot Runners

Flow Channel Design

Hot runners are mounted between the extruder and the injection cavities. Their function is to distribute the single melt stream into the appropriate number of streams depending on the cavitation of the tool. Each cavity should get the exact same amount of material and at the same speed and

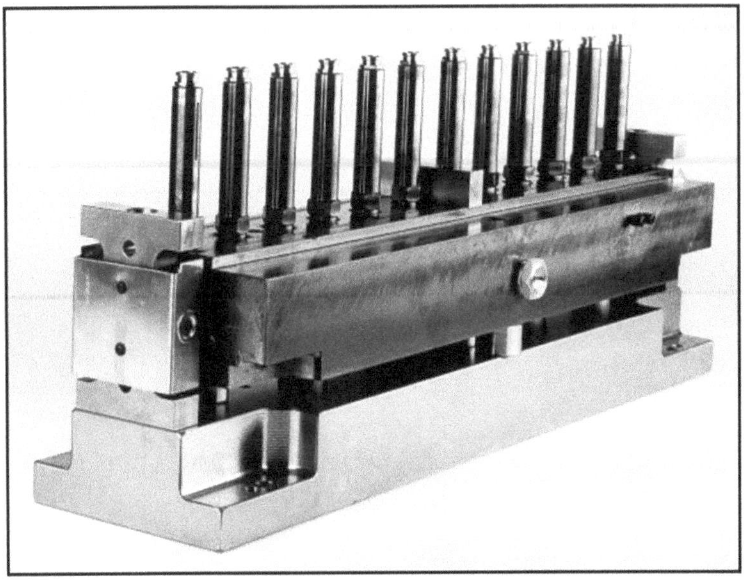

Figure 8.13 All hot runners in single-stage molding like this 12-cavity model feature all nozzles in one or two row(s). Diagram courtesy of Synventive Molding Solutions.

pressure. This is best achieved by "naturally" balancing the flow channels. This means that the path from the extruder to each cavity has the same length and number of turns. This is relatively easy for cavitations whose numbers are a power of 2, that is, 2, 4, 8, 16, and 32. However, even with this condition in place, there are imbalances in any hot runner system that stem from the uneven heating up of the melt stream commonly referred to as viscous heating (Fig. 8.13).

In order to understand this process we have to examine what happens inside the molten resin during injection. PET flows through the barrel, hot runner channel, and nozzles in a laminar fashion not unlike honey flowing through a squeeze bottle nozzle. This flow is characterized by the highest shear rates occurring at the channel walls while there is much less shear in the center of the channel that features the highest speed. There is also no shear in the areas directly adjacent to the runner walls as the speed drop to zero (Fig. 8.14).

Shear deformation causes internal friction between adjacent entangled polymer chains, which results in shear heating. As a result of the laminar flow there is an elevated temperature in the ring-shaped area just off the channel wall (Fig. 8.15).

Most hot runners for single-stage machine are designed in a way that one large channel diverts into two smaller channels that come in at 90 degree.

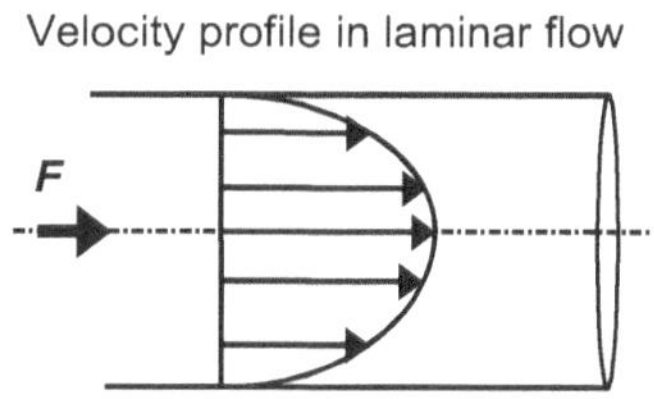

Figure 8.14 Friction just off the channel walls causes viscous heating.

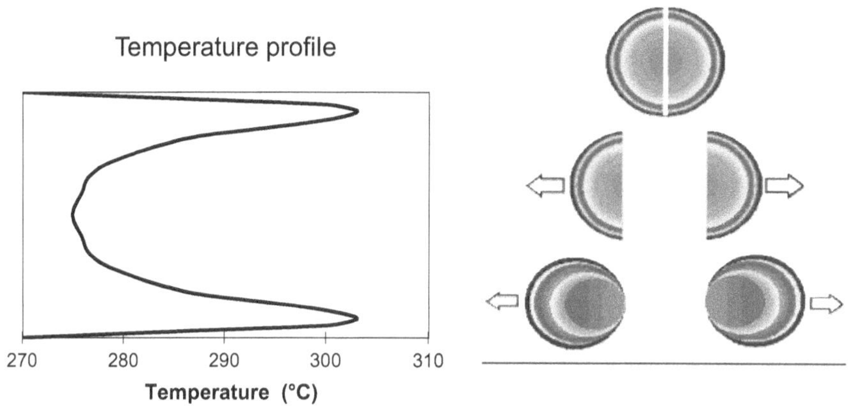

Figure 8.15 Ring-shaped temperature spike in melt channel.

As you can see, there is a tendency for the cooler material from the center of the channel to flow toward the far side of the intersecting channel whereas the hotter material flows to the near side. While intermixing and subsequent temperature homogenization is happening in many cases (depending on hot runners used, material properties, temperature, pressure, and speed as well as shot weight) not enough to prevent the hotter resin from being injected into the side of the cavity that faces the back of the machine whereas cooler material is injected in the front. Because PET is very sensitive to even small temperature fluctuations processors often fight uneven bottle walls with thinner sides toward the parts adjacent to the back of the machine (Fig. 8.16).

Uneven heat of the melt also causes the melt to move at different speeds into the cavities. This can be established by doing a so-called "short shot" where an insufficient amount of material is injected (or the injection is interrupted at some point). It can then be clearly seen if all cavities are filling at the same speed. Uneven filling means that the heat profile in each preform is slightly different which always results in wall thickness differences of the blown bottles (Fig. 8.17).

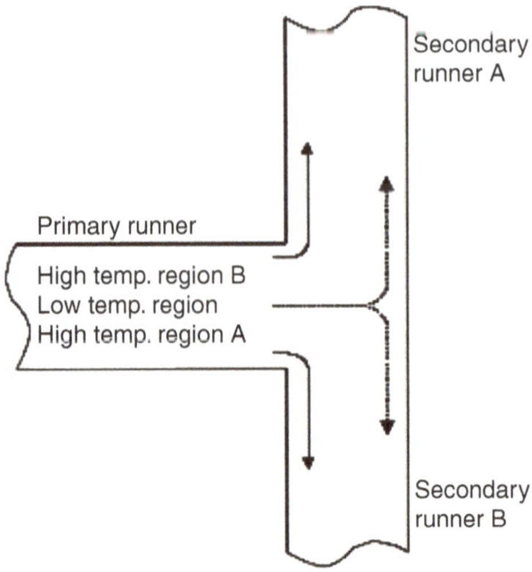

Figure 8.16 Melt distribution in typical single-stage hot runner setup. Hotter material may be pushed to the near side of the secondary runner.

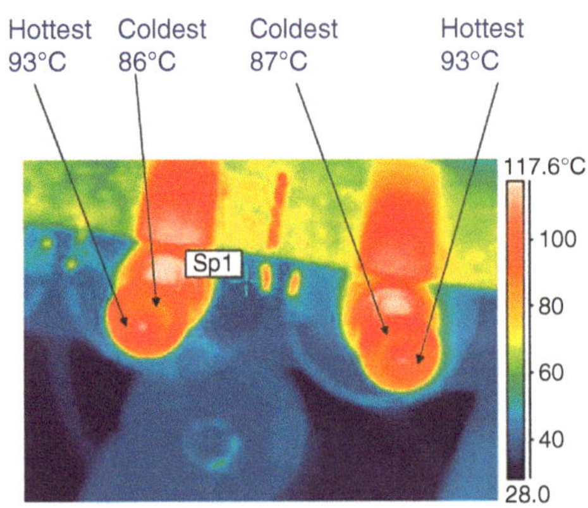

Figure 8.17 Preform temperatures. Actual infrared photograph of preforms after injection. Large side-to-side temperature variation is clearly visible.

However,there is a way to improve on some of the inadequacies of this process. Uneven filling is also a problem in two-stage injection molding. While it does not affect bottle properties, uneven filling and with it cooling can lead to gate and packing problems on some cavities. This is because the material in the cavities that are filled first cools down quicker and increases in viscosity earlier compared to the late-filling material. This may lead to a situation where the fast-filling cavities are overpacked and/or the slow-filling ones are underpacked. The preforms from the former may be

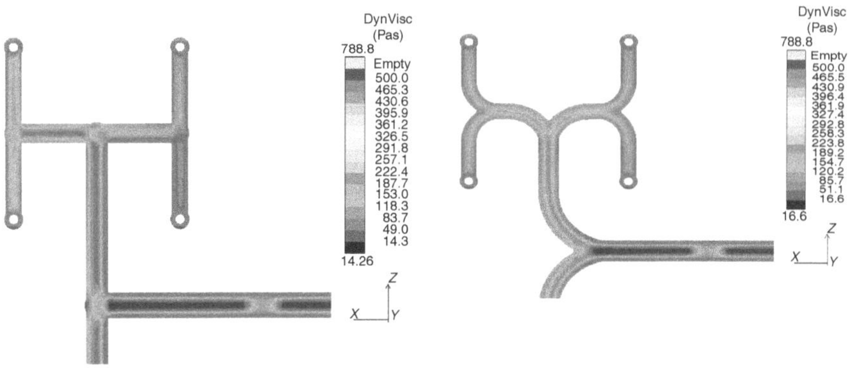

Figure 8.18 Conventional balance (left) and curved balance (right). Lower temperature variations and with it more equal flow are the advantages of brazed hot runner plates. Diagrams courtesy of Mold-Masters.

hard to blow, preforms of the latter could show flow lines or sink marks. The degree to which this happens and how much preforms will be affected depends on the particular preform geometry and process conditions but will be a factor when the fastest cycle time is the primary goal.

One way to improve on this is to change the way the melt changes direction in the hot runner. Instead of changing in hard right angles a brazed, rather than drilled hot runner plate allows the gentle curving of the directional changes leading to less of a melt separation by temperature. In brazing the hot runner plates, two individually milled plates are fused together after milling allowing curved melt paths (Fig. 8.18).

For cavitations other than multiples of two, the path width has to be adjusted to achieve a similar result. In practice, these manipulations often work well for a certain gram weight but fail to deliver balanced flow with others. A variety of measures have been employed over the years with varying success. Most hot runners are also not naturally balanced. Natural balance means that the path of the melt to each cavity has the same length and number of turns. Because of geometry, that is, all preforms in one row, this is often not possible and as a result preforms do not fill at the same speed, aggravating the problem. Changing nozzle diameters in such a way as to allow slower moving melt through larger openings is helpful but usually can only be optimized for a very narrow weight range. Another solution is to add obstacles into the flow path of the faster moving cavities. In PET injection molding for the two-stage process, only certain cavitations are used in order to prevent a departure from a naturally balanced design. In most hot runners, the one flow is divided into three and that is the reason for the 48, 72, 144, and so on cavitations. In one-stage just about any cavitation between 1 and 16 has been built. This is because converters want to use a given machine frame to the maximum cavitation for a bottle application. This has forced hot runner manufacturers to come up with different approaches to still deliver good flow behavior.

While PET is very good at self-leveling (the strain-hardening effect that compels initially warmer areas to blow out after cooler areas have blown) is not large enough to mask the uneven preform heat.

Gate Mechanism

There are two ways to separate the hot melt coming through the hot runner from the cooled preform: thermal and mechanical. Thermally gated hot runners are most prevalent in one-stage while it is the other way around in two-stage injection molding.

In a thermally gated hot runner the break point between the hot and cold melts is controlled by temperature alone. As seen in Fig. 8.19 the temperature difference between the cold cavity and the hot melt is substantial at around 255°C (491°F). A suitable insulation made from stainless steel or other material with insulating properties separates the two sections. When the machine ejects the preforms the melt breaks at the point where the cold gate vestige connects with the hotter material inside the nozzle. This material is already partly cooled down and so has a higher than melt viscosity. This prevents it from seeping into the cavity or lead to "stringing," a common defect where small strands of PET from the melt stream are pulled out with the preform during ejection. An air gap may also assist in the separation process.

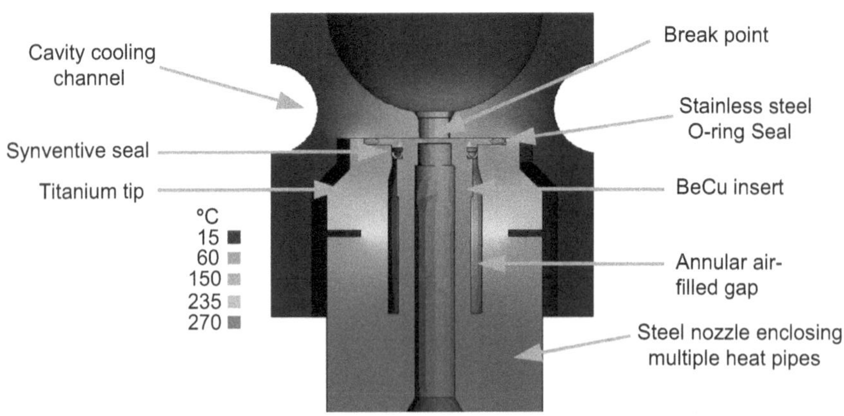

Figure 8.19 Thermally gated hot runner nozzle. The temperature difference between the cold cavity and the hot melt is used to break the preform free off the melt. Diagram courtesy of Synventive Molding Solutions.

Valve-gated hot runners use a mechanical seal between the hot and cold areas by means of a pin called the valve stem. Usually around 3 mm (1/8 in.) to 5 mm (0.2 in.) in diameter, this pin moves back allowing melt flow into the cavity, stays back during hold time, then moves forward controlled by a timer that energizes with the end of hold time. Gate vestiges made with valve-gated hot runners are on average shorter than with thermally gated ones and the cut-off is more precise. Properly operated they also tend to give less problems with stringing and gate crystallinity. Disadvantage of valve gates is the higher maintenance requirement for the pins and the air cylinders driving them (Figs. 8.20 and 8.21).

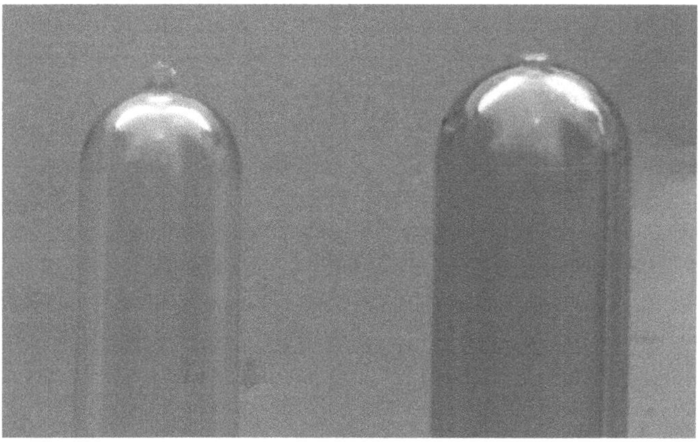

Figure 8.20 Typical difference in length of gate vestige between preforms of thermally gated (left) and valve-gated (right) hot runner.

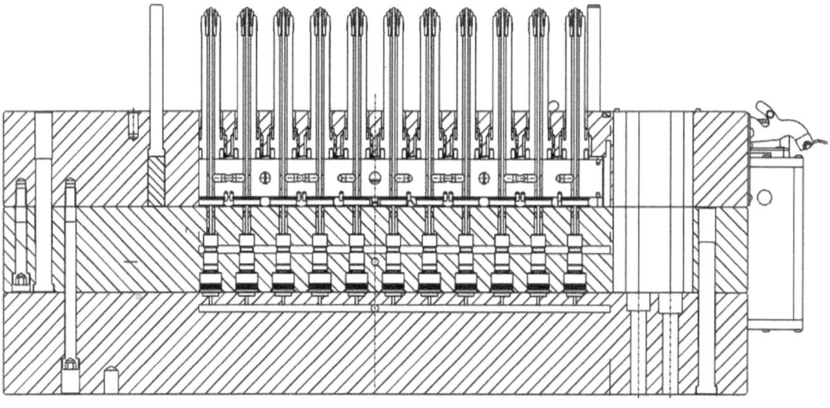

Figure 8.21 Valve-gated hot runners offer more precise control over the gate vestige of preforms. Drawing courtesy of Mold-Masters.

8.8 Integrated Two-Stage Stretch Blow Molding

This process is sometimes referred to as "one-and-a-half stretch blow molding." It is similar to one-stage molding in that both preforms and bottles are made in the same machine. It is also similar to two-stage molding as preforms are completely removed from the injection tooling, placed on mandrels, heat-conditioned in some way, and then blown. The impetus for this development was the above-mentioned fact that the blow stations in one-stage machines are underused. The concept of ITSBM machines then is to reduce the number of blow cavities in some ratio to the number of injection cavities. The blow part of the machine cycles is two or three times for every injection cycle and has only half or a third blow cavities, respectively as compared with the injection cavities. For custom molders with small-volume jobs this can result in significant savings as less blow cavities have to be purchased. It also speeds up the machine because fewer parts have to move at the same time. It adds complexity though as a robot of some sort is needed to take preforms out of the injection tool, turn them 180 degrees, and then place them on mandrels. Machines as large as 80 injection and 40 blow cavities and outputs of up to 36,000 bottles/h have been built but more common are machines with 6 injection/2 blow or 16 injection/8 blow cavities (Fig. 8.22).

Heat conditioning preforms on mandrels can be done in several ways and infrared ovens like the ones used in two-stage stretch blow molding are often used. Another method is to blow hot air against cooler parts of the

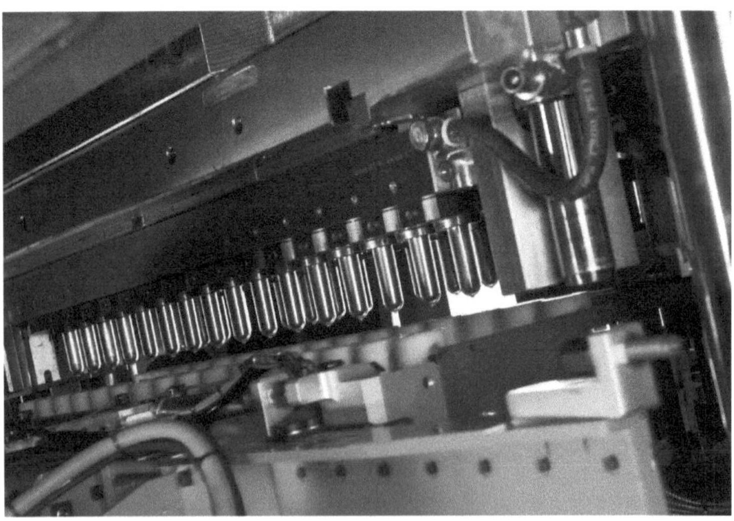

Figure 8.22 Thirty two preforms are being transferred to mandrels on this machine. Picture courtesy of SIPA.

spinning preform. This is especially useful to heat up material just underneath the transfer ring. This part of the preform tends to be cooler than the area around the gate as it had longer time to cool during injection. This is much more effective than the heater bands that are used for the same function in standard single-stage machines.

ITSBM machines share most of the same advantages and disadvantages as one-stage machines but do not have automatic neck orientation as they do not keep the preforms in place. They also add another complication that may sometimes cause problems: the two parts of the machine (injection and blow) need to be synchronized in some way, which may limit options for the processor. Given a certain injection time, blow timing has to be limited to a value that allows both parts to stay in synchronized and vice versa. In most cases this is not an issue with machines that feature 2:1 ratios but may be more challenging when a 3:1 ratio is employed. Whenever the total time of the blow time cycles (2 or 3) exceed the injection cycle, the injection part of the machine has to add the missing time somewhere. This may be by adding cooling or mold open time. However, this changes the preform temperature and so affects the blow process. It is important that the machine warns the operator of this condition so that he/she can adjust the process accordingly.

8.9 Single or Two Stage—That is the Question

At the end of this chapter, I want to discuss the question many small and medium applications beg. Both processes have their distinct advantages and disadvantages and brand owners are well advised to know those in order to make the right choice for their products.

The advantages of single stage are repeated as follows:

- blemish-free bottles
- transfer ring not necessary
- control over preform production
- good conditioning possibilities for oblong bottles on some machines
- thread start can be chosen to coincide with bottle shape
- compact and flexible

This makes this process a shoe-in for all nonbeverage containers. However, there are also some disadvantages:

- long cycle times
- long changeover times
- uneven wall distribution

- quality problems with thermogated hot runners, valve-gated ones are available and are recommended
- machines should be run 24 h/day to avoid higher scrap percentage
- longer learning curve for operators as two processes and PET drying must be mastered
- inefficient blow station as the injection station always has precedent over the cycle time

The latter issue has led to the subcategory of integrated two-stage machines that has been described previously. Thermo-gated hot runners are inferior compared to valve-gated ones but most machines still run with the former.

Now let us look at the advantages of the two-stage process:

- scalable from 1,000 to 81,000 bottles/h
- fast cycle times
- fast changeover
- flexibility; preforms can be made elsewhere and stored
- very good wall distribution for round bottles
- on average lower gram weights are possible for round bottles
- process can be stopped at any time

The main disadvantage is preform damage that occurs when preform tumble onto conveyor belts and storage containers and then again when they are dumped into the hoppers of blow molding machines. Many of the little nicks and scratches can be stretched out when high stretch ratios are used but this is not always the case especially when a preform is chosen from a vendor and does not exactly fit the bottle to be blown. Wrap-around labels or sleeves are a good way to hide these marks and have added to their popularity.

A lesser known problem with small-cavity machines is that heat to each preform can be quite different when indexing machines are used (Chapter 3, Section 3.3). Rotary machines of course do not have this problem at all as each preform gets identical heat.

When we examine the PET market we find that over 80% of all bottles produced are for beverage and the vast majority is in two stage. It is the remaining 15–20% where either process is an option. Many decisions are actually driven by tool prices. Buying even a four-cavity single-stage tool can be hard to justify for volumes of under 2 million/year, which many of the custom applications are. There is less capital expense buying preforms in and run them on a two-cavity reheat machine. The number of available preforms has increased dramatically over the last 10 years with vendors spanning the globe as preforms, unlike bottles, are cost-effective to ship.

In order to make an informed choice which process to use one needs to understand the particulars of the application and determine either on the basis of bottle features or economics. If blemish-free, oblong bottle shape, fixed thread start all come together single stage is an easy answer. If none of those applies the economics must be scrutinized. The preforms for single stage are always custom-made (unless you plan to make a variety of shapes out of the same preform) so the fit is guaranteed. If preforms are bought however, the quest for the right one starts. Not every preform that has the right neck finish and weight is suitable for a particular bottle. An expert should evaluate the available preforms and choose the best fit.

Let us look at a hypothetical application. The container is a round 1-L bottle with a 33-mm neck finish and the end customer wants a weight of 42 g. Yearly volume is 750,000 bottles.

In two stage this will require a one- or two-cavity machine that will probably run around 600–800 bottles/cavity per h. While machine manufacturers publish much higher numbers most custom applications run slower for a number of reasons. This will keep the one-cavity machine busy for 1152 h and the two-cavity machine for half of that. Capital cost is quite low as only one or two blow cavities have to be purchased.

In single-stage machine cycle time will be about 13–16 s, let us say 250 cycles/h. A two-cavity system would then make the required bottles in 1500 h, a four-cavity system in 750 h. Capital cost is significant for each cavity and of course most machines will be able to run four or even six cavities. Here the decision would be based on cost per piece in different cavitations and what future outlook the job holds.

In summary, the decision which process to use can be quite complex and I hope I have given you some pointers that will allow you to choose the right process for your application.

9 Special Applications

Chapter Outline

Stretch Blow Molding. http://dx.doi.org/10.1016/B978-0-323-46177-1.00009-3

9.1 Simulation of the Blow Process

Processors have tried for many years to visualize what happens inside a blow mold during the blow process. With the advent of finite element analysis and ever-faster personal computers, it has now become possible to simulate this event at a reasonable cost and with some degree of accuracy.

Advanced software uses a viscoelastic model that takes into account the dependence of the stress on the deformation history. The program calculates both axial and hoop stress under the impact of the stretch rod and blow air to a high degree of accuracy. It takes into account temperature dependency as well as heat transfer and friction between the different parts involved.

Preparations

Users are required to enter five components to start a simulation:

1. 3D mold or bottle data (best obtained from the mold maker). If bottle data is used shrinkage allowance must be made.

2. Preform data entered as a simple text file. Only dimensions below the neck support ring need to be entered.
3. Stretch rod dimensions, entered by modifying a sample file provided.
4. Stress/strain data of the resin used (supplied by either the material suppliers or independent labs).
5. Process data relating to timing of stretch rod and blow air.

While physical dimensions of preform and mold are straightforward, temperature data are more complex. Some programs offer simulations of the reheat process which vary with lamp settings and cooling air volume. However, margins of error are relatively large since lamp and reflector wear affect heat output, and air volume may be difficult to calculate exactly as it affects preforms. Temperatures may instead be measured with the infrared sensor on the machine (moved to several locations on the preform) or a thermo-camera. Even then the data so obtained is not totally reliable since the surface temperature measured does not tell us the temperature further down in the preform wall. The proximity of the reading to the actual temperatures as an average through the wall thickness (which is the value needed) will depend on equilibration time after the oven section. With some experimentation an adjusting figure may be devised to bring the measured temperature closer to the average one.

The process parameters cannot be taken at face value from the machine operator interface. For example, there is a delay between the blow air valve switching and air actually entering the preform. It will depend on how fast the valve is moving and its distance from the preform. For these reason the values need to be fine-tuned to reflect what is happening inside the blow mold. Some machine controllers provide detailed information on pressure build-up and stretch rod movement inside the mold and the values recorded there should be used. Otherwise, speed and pressure transducers may be mounted to record the exact timing of stretch rod and blow air (Figs. 9.1 and 9.2).

Simulation and Output

Once all data have been entered simulation may start. It may take 5 min to several hours for a simulation to complete. Here are the factors affecting simulation time:

- The software algorithm itself.
- The number of nodes in the preform. Higher numbers result in better resolution but carry a time penalty. Some programs offer the ability to create half or even quarter preforms for symmetrical bottle shapes.

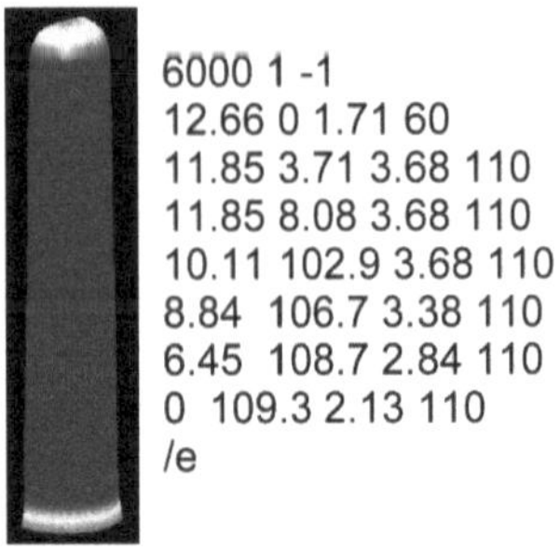

Figure 9.1 Some programs allow preform creation from simple text files. Here a half-round preform with the text used to create it. Sample taken from B-Sim simulation software.

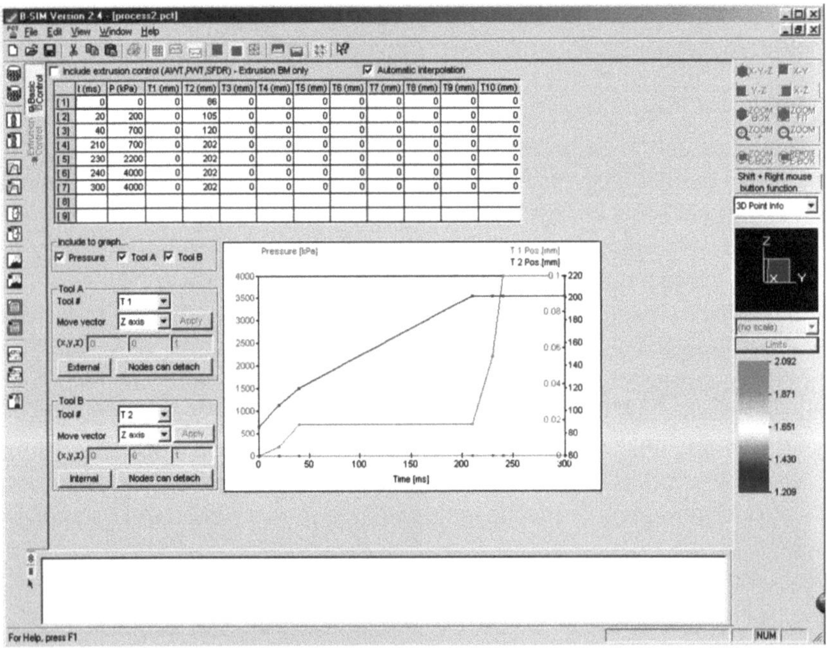

Figure 9.2 All relevant blow parameters can be programmed into the simulation software. Sample taken from B-Sim simulation software.

- The size of the mold date file. If bottle geometry is not too complex it is often advantageous to create a simplified mold model rather than taking the often large mold maker file that carries a lot information unnecessary for the simulation.
- Type of computer used in the simulation. Pentium 3 machines above 1 GHz or equivalent are suitable. RAM should

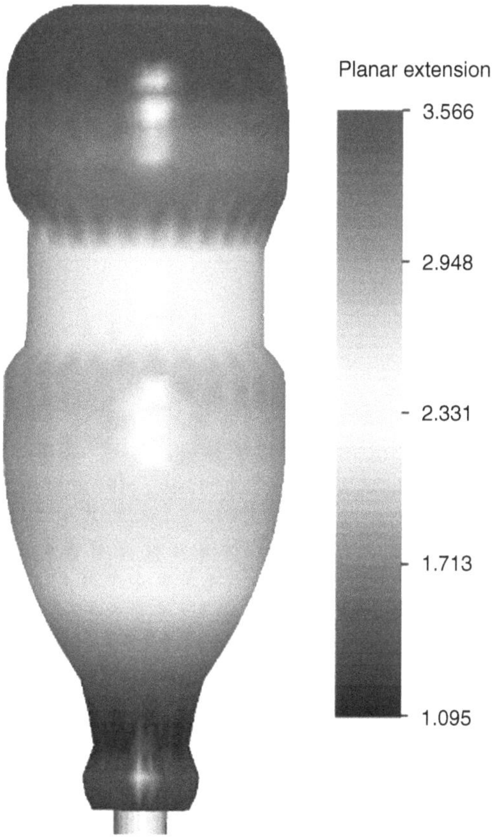

Figure 9.3 Programs yield bottles with a variety of data. Here is an example of planar extension. Sample taken from B-Sim simulation software.

be 500 MB or higher. Each simulation may require 5–50 MB of hard disk space.

After simulation bottles may be examined for temperature, wall thickness, stress concentrations, and planar extension. Some programs may offer modules for postmold cooling, and venting of critical areas. Batch processing may also be available (Fig. 9.3).

Applications

After testing existing preform/bottle combinations the assembled knowledge may be transferred to new projects. There are a number of engineering tasks where simulation software can be of help:

- Preform design: An unlimited number of preform configurations can be evaluated for wall thickness distribution, etc.
- New bottle designs can be tested against existing preforms or to investigate how well new features will be accepted by the process.
- Time to market may be reduced by:
  - faster preform development,
  - easier machine adjustment with parameters taken from the simulation software, and
  - eliminating time-consuming and costly tool rework.

Costs and Benefits

There are other costs in addition to the purchase price of the software. Training of an experienced processor will take some time, during which he or she will be unable to work on resolving production problems. There is also a considerable cost involved in acquiring machine parameters in real time for input into the software. This might require stopping production machinery to mount transducers and taking readings under various process conditions.

Benefits can be significant and easily outweighing costs. Presently most companies go through a very cumbersome process when starting up a new bottle. Single-cavity development is still a common method of avoiding production tool rework cost. Setting up the single cavity, and later the production machine, is a time-consuming task. Simulation software may be tuned to provide stable data output allowing the elimination of the single-cavity tool as well as giving processors a set of machine parameters that they can use to start the production machine. Fine-tuning will always be necessary as no two machines are alike but simulation software might be able to greatly simplify this task.

9.2 Stretch Blow Molding of Oriented Polypropylene

Oriented polypropylene (OPP) belongs to a group of thermoplastics called polyolefins and is related to materials such as high-density polyethylene (HDPE) and low-density polyethylene (LDPE) besides standard polypropylene (PP).

The history of stretch blow molding OPP started in the 1960s when the Beloit Company reheated extruded PP pipe to form and blow biaxially oriented containers. Over the years resin suppliers have improved material

Table 9.1 Differences Between OPP and PET

Parameters	Opp	Pet	Comment
Specific gravity (g/cm^3)	0.90	1.35	OPP is 33% lighter
Cost (US$/kg)	$1.45	$1.45	Resins are currently at par
Hot fill capability (°C)	90	50	Standard OPP can be hot filled
Water vapor transmission rate (g/m^2 × 24 h)	0.29	2.3	OPP 8 times better
Oxygen permeation (cm^3/m^2 × 24 h × atm)	73	2.2	PET 30 times better
CO_2 retention (cm^3/m^2 × 24 h × atm)	129	14	PET 9 times better
Typical preform molding cycle	18 s	12 s	OPP molding 50% slower
Process window (°C)	±2°	±8°	OPP is harder to process
Tensile strength (MPa)	33	57	PET is almost twice as strong
Maximal stretch ratio	8	16	PET can be stretched further
Drying required	No	Yes	Cost savings with OPP
AA creation	No	Yes	No taste changes over time with OPP

characteristics to the point where the stretched portions of OPP are about as clear as polyethylene terephthalate (PET) whereas unstretched portions such as the neck and parts of the bottom stay "milky." Since these areas are mostly not visible to the consumer at the point of purchase, there is virtually no difference in appearance between bottles made from the two resins.

There are several distinct differences between OPP and PET (Table 9.1).

The poor mechanical strength of OPP has frustrated users that try to take advantage of OPP's lower density. At the same wall thickness (and therefore 33% less weight) top load performance of OPP bottles is significantly lower and OPP bottles in the marketplace are about at the same, if not slightly higher weight than PET bottles of the same volume as thicker walls are needed to make up for the lower mechanical strength.

Process Difference

OPP does not strain harden like PET. One effect of strain hardening is that weaker or warmer preform areas will stretch first but as strain hardening occurs these areas become harder to stretch thereby allowing surrounding areas to stretch. This has the effect that thin spots are minimized and it is not unusual for a round PET bottle to show wall thickness variations in one circumferential spot of less than the 0.02 mm (0.001 in.). This is not the case with all polyolefins. A thin or warm spot in the preform will always translate into a thin spot in the bottle making the achievement of minimum wall thickness in sensitive areas such as bottom corners more of a challenge.

OPP does not have the same absorption properties as PET. While it is beneficial for PET to keep lamp settings at values over 80%, the inner wall of an OPP preform heats up better when lamps are dialed down to 45–55%. Considering that OPP also needs to be reheated 30°C more than PET this of course requires a larger number of ovens for OPP reheating, which forces processors to significantly reduce machine output when they process a OPP preform on a standard blow machine. A machine to handle all OPP preforms should have approximately twice the number of ovens than cavities on the same machine.

A longer blow cycle is often required because OPP cools down slower not just in the injection but also in the blow process.

The small process window causes the OPP process to be quite sensitive in day/night cycles in plant and chiller water temperature and air conditioning the blow machine is strongly recommended to avoid large scrap percentages.

In order to get good clarity an axial stretch ration larger than is common for PET is recommended, up to 3:1 is not unusual. This in turn leads to shorter, stubbier preforms compared with PET.

Applications

From the date mentioned previously there seem to be three applications for OPP:

1. Water bottles
2. Pharmaceutical, personal care, and household cleaner bottles
3. Hot-fill drinks with a maximum filling temperature of 90°C

A major advantage of OPP for water bottles is that no acetaldehyde (AA) is created. AA is a naturally occurring sweetener that is a by-product of PET processing. While harmless to consumers, it must be kept to very low levels so as not to impart a taste to bottled water. Improper drying of PET is one of the main reasons for process and bottle performance problems. Eliminating the drying altogether allows OPP processing at higher operating efficiencies.

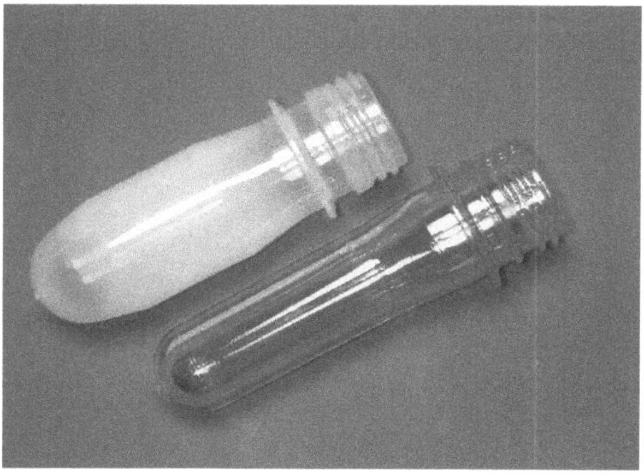

Figure 9.4 OPP and PET preforms for the same bottle. The larger diameter of the OPP preform is necessary to accommodate OPP's lower maximal stretch ratio. Photo courtesy of Bekum America Corporation.

OPP has not made any major inroads into this market. However, its lower tensile strength requires the OPP water bottle to be thicker than its PET counterpart thus reducing or eliminating the density advantage. The longer injection molding cycle time also adds cost to the OPP bottle (Fig. 9.4).

Furthermore, because of its poor oxygen barrier gases (like diesel fumes) can actually enter the bottle causing taste changes in the product.

The second category of applications is more promising. Especially in pill bottles the better water vapor transmission is a plus to keep medicine dry over long periods of time.

The third set of applications held the greatest promise of OPP bottles. As explained in Chapter 10 PET bottles for hot-fill applications are rather expensive and OPP seemed to be perfectly suited. However, practically all hot-filled products are also oxygen sensitive and it is OPP's poor oxygen barrier that has hindered its use (Fig. 9.5).

Multilayer

Using ethylene vinyl alcohol (EVOH) or nylon as a barrier layer can alleviate OPP's poor permeation characteristic to oxygen and carbonation. To bond the barrier material to the OPP two adhesive layers were needed thus creating a 5-layer preform or even 6-layer when used in the extrusion blow molding process. This would require very sophisticated hot runner systems for injection molding and is easier accomplished by extrusion blow molding. A patented process called gamma-clear is now being marketed that allows the before impossible fusion of OPP with a barrier

Figure 9.5 Hot-fill bottle in OPP. OPP's poor oxygen barrier has so far kept it from becoming a major resin for this application. Photo courtesy of Bekum America Corporation.

without the use of an adhesive layer. However, care must be taken to avoid delamination that can occur during the stretching of the preform.

In the extrusion blow molding process only wheel machines produce enough preforms to get the same economics that are prevalent in injection molding of PET preforms. Wheel machines are in use today for molding multilayer, extrusion blow molded 8 oz juice and ketchup bottles that compete with PET bottles. They may be lower in cost (depending on output) but lack the clarity that only OPP can provide (Fig. 9.6).

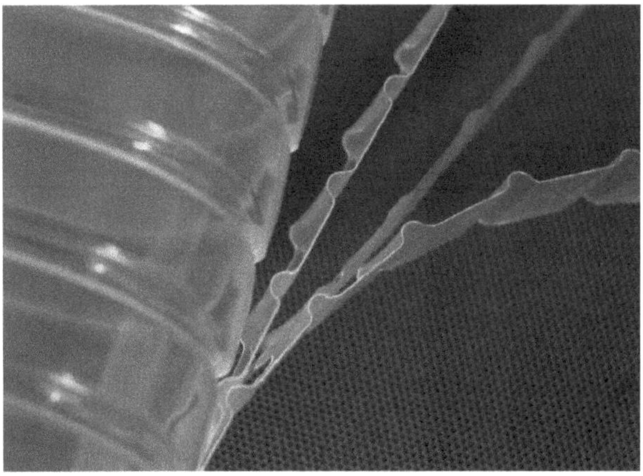

Figure 9.6 Multilayer construction OPP/EVOH/OPP. Photo courtesy of Bekum America Corporation.

Will OPP replace PET as the material of choice for many packaging applications? That is unlikely to happen but OPP can expand on the current niche and venture into areas where its properties such as lower overall cost, hot-fill capability, and superior water vapor barrier will make a very competitive package. Economical preform production is a major obstacle that could be overcome. Barrier technologies have to prove to be cost effective and guaranteeing the barrier performance over time.

9.3 Plant-Based Plastics

Increased concerns over our rapidly diminishing fossil fuel reserves as well as the piling up of plastic bottles in landfills have led to the development of plant-based resins (bioplastics) of which PLA (polylactic acid) is the most prominent. From the beginning, there has been controversy on how much more energy efficient plant-based resins can be produced, recycling issues, composting questions, and the problematic conversion of arable land from food to plastic resource. While I cannot delve into this discussion in greater detail, I'd like to propose a few points for consideration.

In terms of energy efficiency PLA and other materials shares similar concerns than glass. While both materials do not require fossil fuel feedstock, their processing and, in the case of PLA the growth of its feedstock, consume considerable amounts of these. In all studies PLA fares well in terms of energy efficiency even though the numbers differ depending on the methodology used. However, all plant-based plastics add to eutrophication, the addition of artificial substances such as phosphates and nitrates to aquatic systems, a well-known problem of industrial agriculture. There are other issues such as emission of carcinogens and ecotoxicity, where bioplastics fare worse than any plastics.[a] Some of these concerns can be mitigated if bioplastics could be made from agricultural waste products such as peels and husks, and efforts are underway to do just that.

Also, there is the misconception of many consumers that assume that bottles produced with bioplastics are biodegradable, meaning they will decompose whether they are placed in a dustbin or the landfill. However, PLA and many other bioplastics are only compostable, requiring elevated temperatures for several days in order to break down, which can only be achieved in an industrial composting facility. If PLA bottles end up in a landfill, their decomposition is not that different from plastics and there are only a limited number of composting facilities in operation. It is also questionable whether it makes sense energywise to just throw these bottles away instead of getting at least some of the energy back. To date, the process to reconstitute PLA into its primary components is just in its infant stage.

Bottles made from bioplastics look like PET bottles and end up in the PET recycling stream where they can cause havoc because of their lower melting temperatures, that is, they burn when processed for PET. As long as bioplastics only come in trace amounts the resulting haze will be not noticeable but at higher concentrations bioplastics will have to be sorted out. Most recyclers use X-ray scanners that work well for PVC but will have to switch to infrared sensors to be able to detect bioplastics at considerable cost to them. Recyclers will end up with a smaller recycling stream once bioplastics bottles are in wide use.

There are now methods to produce ethylene from plant-ethanol. Ethylene is the precursor to ethylene glycol, one of the two main components of PET manufacture. Bottles produced from this resin then feature 30% plant-based resin by weight, somewhat a lesser percentage in carbon emission as the other component of PET manufacture, terephthalic acid, is higher in carbon emissions. Coca Cola has started this approach and is experiencing no issues in preform manufacturing and many of their bottles are featuring a logo stating the 30% natural content. Both Coca Cola and Pepsi Cola seem to avoid getting caught in the reported recycling problems of PLA. Instead, the goal is to change the feedstock of the two main components of PET and create virgin resin that is not distinguishable from PET made from fossil fuel feedstock.

---

[a]Sustainability Metrics: Life Cycle Assessment and Green Design in Polymers, Environmental Science and Technology, Sep. 2010.

In short, while bioplastics are well received on the consumer level the industry would do well (and some companies have done so already) to study all related concerns before aggravating environmental or recycling issues. If fossil fuel prices increase again from current levels prices for bioplastics will rise less and will become more attractive to brand owners. This could increase their use significantly.

Processing PLA

Since PLA is the most prominent and best studied of the bioplastics, I'll concentrate on this material for the discussion (Table 9.2).

PLA melts at a temperature of around 200°C (392°F). Since it is so similar to PET users may be tempted to use existing PET tooling for PLA production. This may not work in all cases because hot runner tolerances are calculated for the higher PET temperatures and some hot runners may leak

Table 9.2 Overview of the Main Characteristics

Parameters	PLA	PET	Comment
Specific gravity (g/cm^3)	1.24	1.35	PLA is very similar
Cost (US$/kg)	$2?	$1.45	PLA price not confirmed
Hot fill capability	40°	50°	Both can be heat set though
Water vapor transmission rate (g/m^2 × 24 h)	20	2.3–3.4	PET ~7 times better
Oxygen permeation (cm^3/m^2 × 24 h × atm)	40	2.2–2.9	PET ~16 times better
CO_2 retention (cm^3/m^2 × 24 h × atm)	200	14	PET ~14 times better
Typical preform molding cycle (s)	14	12	PLA and PET are similar
Process window (°C)	±5	8	PLA and PET are similar
Maximal stretch ratio	11	16	PET can be stretched further
Drying required	4 h@80°C	4 h@160°C	Drying of PLA less expensive
AA creation	No	Yes	

because of the lack of expansion. PLA needs to be purged with PP before and after use and users should implement a strict procedure to avoid problems. Mold temperatures are slightly higher compared to PET, between 20 and 40°C (68 and 104°F). The injection capacity should be a close match to the preform output as prolonged residence time (as well as excessive shear) tends to degrade the material. Standard PET screws work well with PLA because of their low-shear design. Preform bases should be kept at a low wall thickness as they are difficult to stretch and tend to need extra cooling time.[b]

In blow molding, stretch ratios are slightly lower to accommodate the lower overall stretch ability of PLA. Paramount to successful processing is excellent oven ventilation. When the ambient oven temperature exceeds 85°C (185°F) control over the preform temperature profile is quickly lost and bottle quality severely compromised. Users that have machines that do not have a thermocouple in the oven section installed should retrofit one even if it means to go with an after-market product as some machine manufacturers do not offer this option. If oven temperature cannot be controlled with existing fans, more/different fans and/or different ducting will become necessary. Although PLA does not absorb heat the same way as PET does a longer soak time, that is, time between end of oven section and start of blowing, is beneficial to raise the inside preform temperature of the preform compared with the outside one.

9.4 Blow Process for Hot-Fill Applications

There are now well over 130 hot-filling lines in North America alone and billions of bottles are produced annually for this process. All noncarbonated drinks containing sugar are sensitive to microbial contamination and therefore require a sterile container. Hot-filling temperatures range from 82°C for juices to 95°C for tomato-based products. To better understand the demands on the bottles a closer look at what happens to the bottle during, and especially after, filling and capping is in order.

The Hot-Fill Process

The product to be filled is flash pasteurized. Depending on its nature it undergoes heating to between 120 and 140°C (248–284° F) for a few seconds followed by a cooling period to the filling temperature. After bottles

[b]Ingeo Biopolymer 7001D, http://www.natureworksllc.com/Technical-Resources/~/media/Technical_Resources/Technical_Data_Sheets/TechnicalDataSheet_7001D_bottles_pdf.pdf

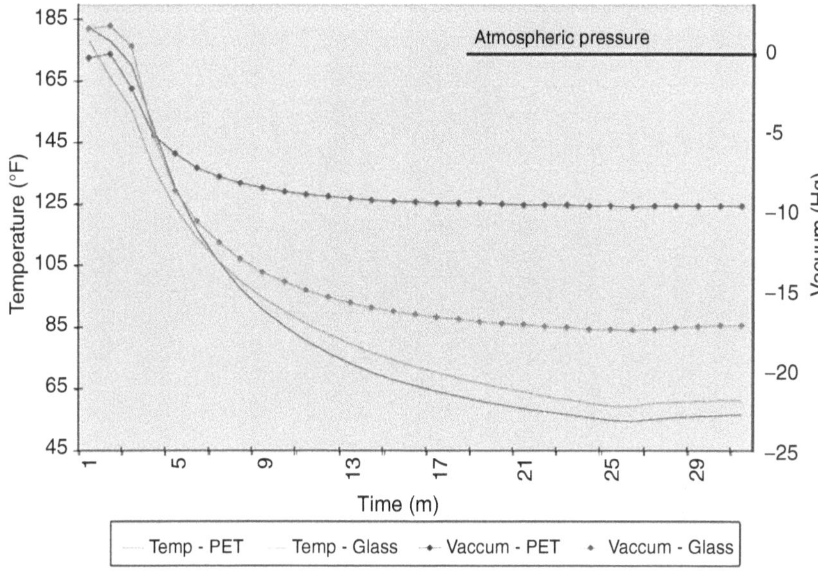

Figure 9.7 While temperatures in a glass and PET bottle behave in similar fashion, the controlled distortion of the PET bottle leads to a smaller vacuum. Diagram courtesy of Amcor Rigid Plastics.

have been filled and capped they are turned upside down or laid on their sides for about 10 s to sterilize the inside of the caps. Afterward they travel for about 20 min through a tunnel where cold-water sprays reduce the bottle and product temperature to below 40°C (104°F).

After capping the internal pressure increases by 0.1– 0.3 bar, caused by an increase in headspace air temperature and shrinking of the bottle due to stress relaxation. (To avoid the former behavior many bottles are filled to the brim without any headspace) (Fig. 9.7).

Once the liquid is at room temperature its volume will have decreased by approximately 3% (depending on the filling temperature). Air solubility in the liquid has increased and caused vapor condensation. Both phenomena result in a vacuum of approximately 0.2–0.3 bar. This vacuum would distort a standard PET bottle, resulting in a very unsightly appearance (Fig. 9.8).

Demands on the Bottle

Bottles must resist the filling temperature and shrink only by a maximum of 1.5–3%. They must have a crystallinity level of above 30% to reduce humidity absorption into the bottle wall, which would otherwise decrease mechanical resistance. High crystallinity levels also increase

Figure 9.8 Vacuum panels contract in a controlled way as the liquid in the bottle shrinks.

thermal resistance, minimizing the risk of bottles distorting. For the same reason regular wall distribution becomes even more important than for conventional bottles as differences in wall thickness around the circumference of the bottle would lead to nonlinear shrinkage. Bottles must also be relieved of residual stress incurred during blow molding. Otherwise they will relieve and distort during filling.

To withstand a significant vacuum bottles have so-called "vacuum panels" and special bases. There are usually six panels spaced evenly around the circumference of the bottle that contract in a controlled manner when a vacuum is created. Bases feature a high dome with, once again, six ribs

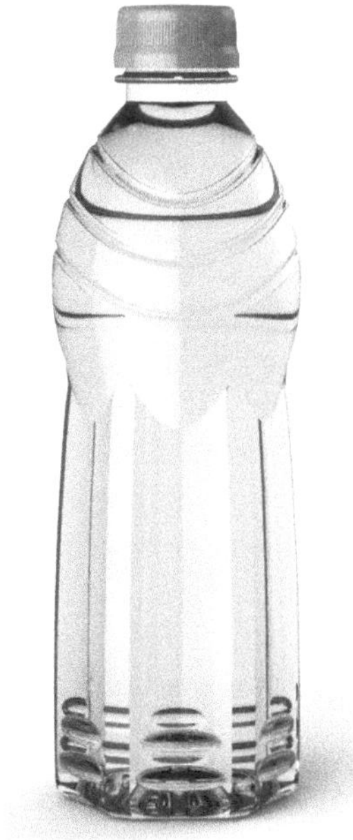

Figure 9.9 Concept drawing for hot-fill bottle with innovative vacuum panels. Target weight of this 20 oz bottle is an astounding 22 g. Photo courtesy of Amcor Rigid Plastics.

protruding to the outside of the bottle allowing the base to shrink upward with no distortion during cooling.

End users favor design freedom and vacuum panels are seen as unsightly, and are usually covered with labels. Converters are now actively working on panel-less designs incorporating other features, such as molded-in grip handles, for the same function or increasing the functionality of the bottle base. Another approach is to add material to the bottle, rendering it rigid enough to withstand the stresses (Fig. 9.9).

Several converters are working on processes to increase the sidewall crystallinity levels to over 40% thus opening the door to PET bottles being used for retort applications. Achieving this crystallinity level is already possible albeit with a severe increase in sidewall cloudiness.

Heat-Set Blow Process

Based on the demands the blowing process for hot-fill bottles exhibit several distinct differences. To withstand the thermal load bottles are significantly heavier when compared with water bottles or CSD bottles.

Container Type 20 oz/500 mL	Weight (g)
Water	8–16
CSD	21–27
Hot-fill	25–36

Depending on the filling temperature as well as size and wall thickness of the container neck, the neck may undergo thermal crystallization, turning white in the process. Thermally crystallized PET is better able to resist thermal distortion and the cap usually covers the white neck so it does not detract from the appearance of the bottle. Neck crystallization is done in special machines that heat neck spinning on mandrels to about 125°C (254°F), then letting them cool down and allowing the crystals to form. As this is an additional and somewhat costly manufacturing step many necks may be designed with thicker-than-normal walls giving them extra stability (Fig. 9.10).

Since bottles need a higher crystallinity level in the sidewalls, but must retain their clear appearance, it is paramount that only crystals that are small enough so as not to refract light are encouraged to grow. Strain-induced crystals measure only 0.5–0.7 μm in diameter and do not refract light in any way.

Figure 9.10 A fully crystallized neck can withstand filling heat better.

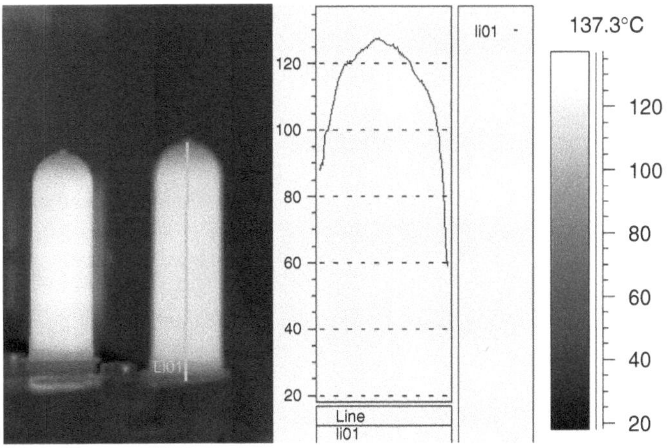

Figure 9.11 Thermo image of preform for heat set at the oven exit. Heat-set preforms are reheated to higher temperatures to encourage crystallization. Photo courtesy of Constar International Inc.

Preforms are reheated to a temperature 10–15°C higher than would be used for the same preform/bottle combination for cold fill. The blow molds are heated to 120–140°C instead of being cooled. Temperatures above 145°C leave a deposit on the bottle walls and clarity is diminished (Fig. 9.11).

When the stretched preform hits the walls of the hot blow mold strain-induced crystals heat up to a temperature conducive to growth. The amount of crystal growth depends on both blow mold and preform temperature and contact time between PET and blow mold walls. Since there is a difference in resin characteristics, product development should test whether higher preform or higher mold temperature leads to lower shrinkage. A typical test would include running preforms with temperatures of 115, 120, and 125°C (239, 248, and 257°F) and cross-reference them with blow mold temperatures of 120, 130, and 140°C (248, 266, and 284°F). A design of experiment with Taguchi methods will reduce the number of tests required (Fig. 9.12).

The base of the container must be kept cool enough to avoid crystallization. This is because the base contains a significant amount of unstretched and therefore amorphous material. This material would crystallize into large, light-refracting crystals thus turning white and brittle. Typical base cooling water temperatures are 30–80°C (86–176°F).

Once the bottle sidewalls have reached a sufficient crystallinity levels, bottles must be cooled to avoid distortion during and after demolding. For this purpose air is introduced through holes in the stretch rods (the

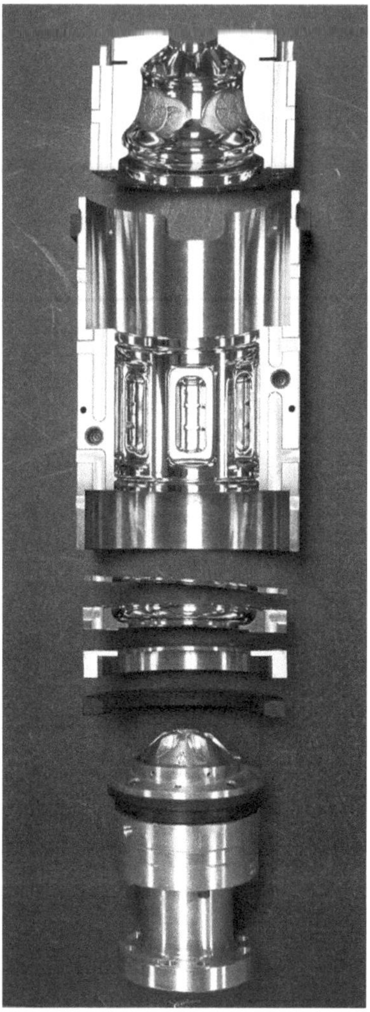

Figure 9.12 Molds for heat set are made from stainless steel. The first of the rings underneath the cavity body is for capacity adjustment. Photo courtesy of Garrtech Inc.

so-called "Balayage" process). This air circulates through the mold cavity for 0.5–1 s, cooling the PET and bringing the crystallization process to a halt. Bottles are then demolded in the same way as other bottles.

Volume Shrinkage Test

Bottles must be tested for volume shrinkage to make sure they will stay at the right volume before filling. Brand owners are always concerned that

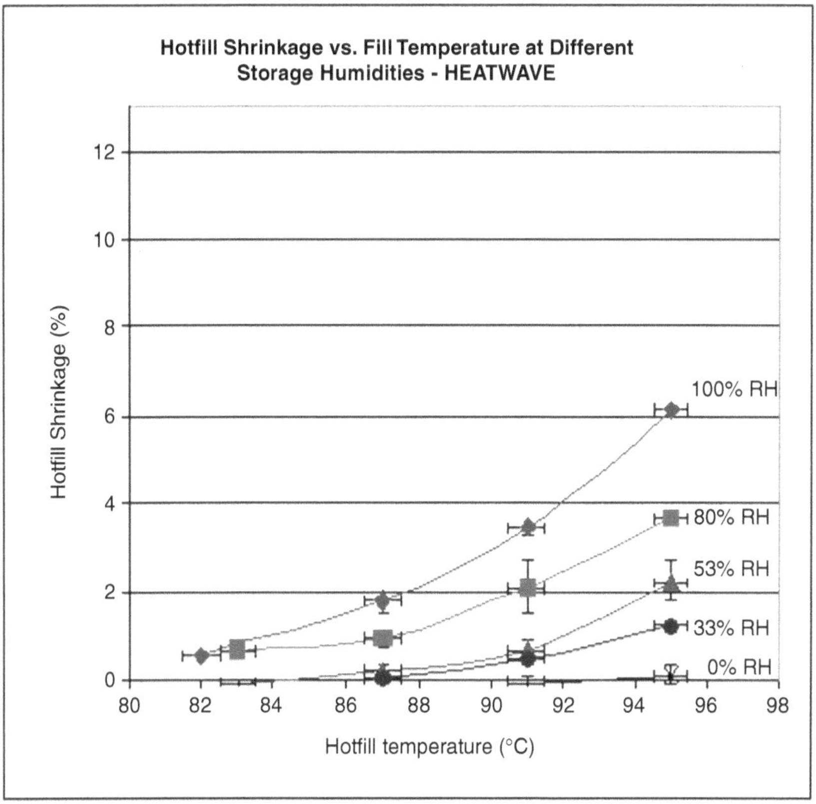

Figure 9.13 Percentage of shrinkage after 30 days of storage under various humidity conditions. Graph courtesy of Eastman.

bottles become too big, the fill level drops, and consumers feel cheated even if the correct product volume is in the bottle. As can been seen from the graph (Fig. 9.13), bottles keep shrinking in storage. Therefore, storage time has to be limited and storage conditions monitored with respect to temperature and especially humidity. The higher these two parameters are, the more the bottles will shrink. If higher shrinkage occurs during storage both brimful capacity and fill level height will be affected and an entire container load may be rejected by the end customer. Typically, there is only a 2% leeway when it comes to volume specification and this can easily be exceeded when production and storage conditions are not adequate. For storage this means the warehouse should be ventilated in high temperature/humidity conditions and containers stored away from south-facing walls and never in trailers. Once the bottles are filled the liquid product acts as a heat sink, limiting further temperature increases that could cause additional shrinkage. At that point barrier properties come into play.

Freshly blown bottles perform much better in this test and should ideally be stored for 48 h before testing. This is of course highly impractical as thousands of bottles will have been produced at the time when the test comes up negative and all these bottles would have to be quarantined and retested. To get around this dilemma, processors wait at least 2 h before testing and then add 2°C (3°F) to the test temperature as a proxy measure that has shown reasonable valid results. Here is the procedure that must be followed.

- Fill a bottle brimful at least 2 h after blowing with cold water and measure the volume. Easiest is to weigh it.
- Dump the water and fill it with hot water at the hot-fill temperature plus 2°C or 3°F.
- Cap it and cool it under cold water to room temperature. Check functioning of the vacuum panels at this point.
- Dump the hot water and cool the bottle with cold water.
- Dump that and fill it with cold water brimful and weigh it again.
- Difference cannot be more than 1.5%, smaller is of course better.

Double-Blow or Two-Wheel Process

When hot-fill temperatures reach around 90°C (194°F) the standard heat-set process runs into difficulties to make suitable bottles that can withstand storing, especially under hot and humid conditions. The standard process can achieve sidewall crystallinity of 32% to a maximum of 35% with a maximum density of 1.37 g/cm^3. It is possible to achieve higher rates but clarity suffers as the bottles turn cloudy. It is also possible to run heat-set machines in-line with filling lines to achieve higher filling temperatures but this is often not practical.

The solution is the so-called double-blow process. Invented in the 1980s it is only now becoming clear what promises it holds especially in the conversion from glass for such difficult products as pickles and salsa. These are filled into wide-mouth containers at temperatures close to the boiling point of water, weigh heavily in the consumer's grocery bags, and will break when dropped from even small heights. They also carry a high carbon footprint because of their weight (hot-fill glass is actually thicker than standard glass), higher transportation costs, and the high temperature necessary for glass production (Fig. 9.14).

The double-blown PET jar on the other side can achieve crystallinity of up to 41% with densities close to 1.39 g/cm^3, making jars that are

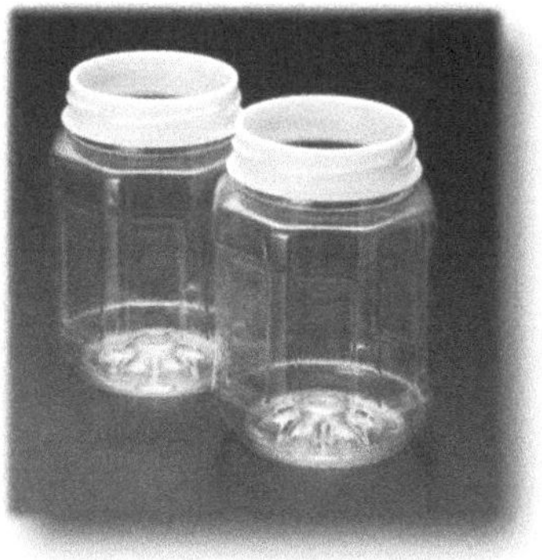

Figure 9.14 Double-blown and heat-set wide-mouth jars are taking over glass jars. Photo courtesy of Nissei ASB.

unbreakable, lightweight, fully recyclable, withstanding steam tunnel pasteurization, and fulfilling all product requirements. However, to achieve this performance a few measures, additional to standard PET jar production, must be taken:

- The neck portion of any preform stays amorphous and is therefore subject to deformation during the filling process. To avoid this, preforms undergo controlled neck crystallization. They are placed on heated mandrels where the neck portion reaches a temperature of about 125°C (257°F). As the PET cools down slowly the molecular chains have time to organize into crystals, which has the effect of the neck to turn white. This results in nearly zero deformation during hot filling (Fig. 9.15).
- Preforms are reheated in the neck down position (Fig. 9.16).
- A special blow machine blows the part in two stages. In stage one the preform is heated to typical temperatures about 100°C (212°F), then stretched and blown into a super-heated bottle mold (up to 200°C or 392°F) with dimensions about 20% larger than those of the finished container. This high temperature, well within the range of optimal crystal growth, encourages crystallization and lead to some cloudiness. After blowing the air is exhausted but the container stays in the mold a short time longer. This relieves incurred stresses and the container shrinks. This process yields bottles that are especially resistant to aging. Aging is a concern with heat-set

Figure 9.15 Necks are heated up to let the amorphous PET in the neck finish crystallize. This stabilizes the neck during the hot-fill process. Diagram courtesy of Nissei ASB.

Figure 9.16 Preforms are reheated to a similar temperature as in the standard process. Diagram courtesy of Nissei ASB.

bottles because stresses tend to relax over time through creep, thus changing bottle dimensions and properties. After the initial blow bottles shrink further back on their way to the final blowing station (Fig. 9.17).

- There the bottles undergo final forming similar to the standard heat-set process with blow mold temperatures in the

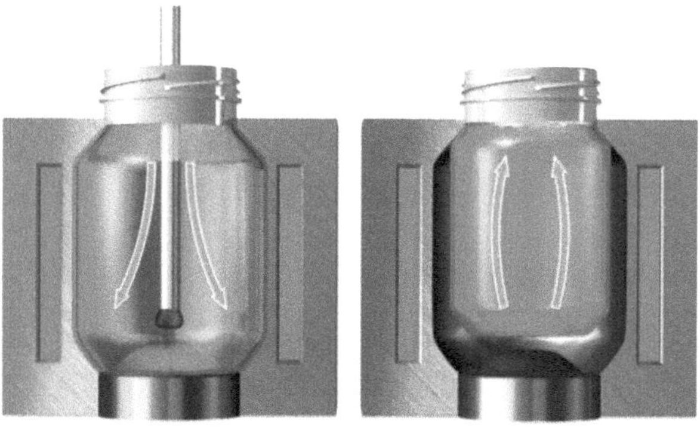

Figure 9.17 The first bottle mold is larger than the final container dimensions and heated to very high temperatures. Diagram courtesy of Nissei ASB.

100°C (212°F) range. This step clears up most of the cloudiness the bottles incurred in the first step (Fig. 9.18).

- A specially designed jar bottom contracts as the vacuum sets in is shown in Fig. 9.19.

A typical glass jar of 350 mL volume weighs about 200 g. The equivalent PET jar weighs 30 g instead.

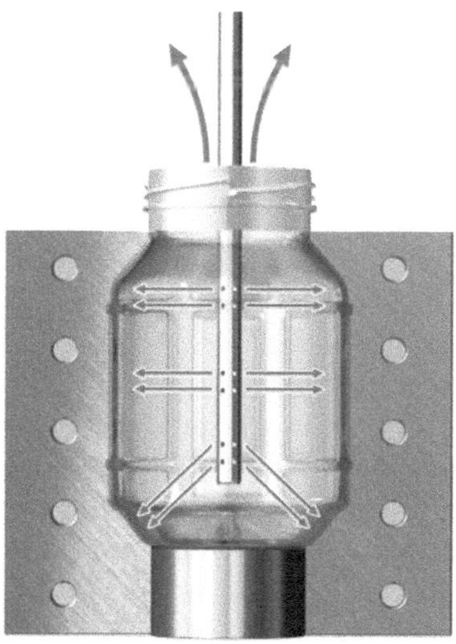

Figure 9.18 The second stage is very much like standard heat setting with cooling air blowing through the stretch rods. Diagram courtesy of Nissei ASB.

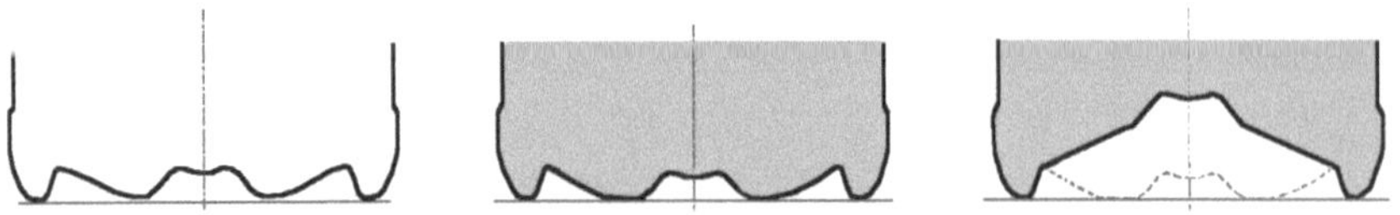

Figure 9.19 The ensuing vacuum in the jar pulls the bottom up.

Economic Considerations

Hot-fill bottles are approximately twice or three times as costly as water bottles. Weights of both declined steadily but the relationship is still the same. The lightest hot-fill bottle is about twice the weight as compared to the lightest water bottle for the same volume (19 vs 8.5). There are five reasons for this:

1. higher material content;
2. increased cycle time needed for crystallization and cooling, dropping outputs to approximately 1500 bottles/cavity per h in the reheat stretch blow molding (RSBM) process for the lower hot-fill temperatures and even more for higher ones (above 90°C);
3. increased (~double) air consumption because of the additional cooling air required;
4. increased cost and maintenance for molds and the frequent cleaning of material residue; and
5. if oil-based thermolators are used for mold temperature control these tend to gum up and need acidic cleaning twice a year as well.

Cold Aseptic Filling

For these reasons cold aseptic filling has become more popular over recent years, especially in Europe. Aseptically filled products are flash pasteurized and then cooled to room temperature in a sterile environment. Standard (lightweight) bottles and caps are sterilized and filled under aseptic conditions. This process allows the product to keep more of its vitamins and natural tastes and is often preferred by consumers. While it requires a very costly and operator-intensive class 100 clean room, near aseptic filling is also popular in a class 10,000 clean room. The clean room numbers refer to the number of airborne particulates.

Table 9.3 shows various specifications.

It depends on the product and desired shelf life and storage conditions which clean room must be used. The savings in bottle costs must be

Table 9.3 Clean Room Specifications

Class	Maximum Particles (m^3)						FED STD 209E Equivalent
	≥0.1 μm	≥0.2 μm	≥0.3 μm	≥0.5 μm	≥1 μm	≥5 μm	
ISO 1	10	2.37	1.02	0.35	0.083	0.0029	
ISO 2	100	23.7	10.2	3.5	0.83	0.029	
ISO 3	1,000	237	102	35	8.3	0.29	Class 1
ISO 4	10,000	2,370	1,020	352	83	2.9	Class 10
ISO 5	100,000	23,700	10,200	3,520	832	29	Class 100
ISO 6	1.0×10^6	237,000	102,000	35,200	8,320	293	Class 1,000
ISO 7	1.0×10^7	2.37×10^6	1,020,000	352,000	83,200	2,930	Class 10,000
ISO 8	1.0×10^8	2.37×10^7	1.02×10^7	3,520,000	832,000	29,300	Class 100,000
ISO 9	1.0×10^9	2.37×10^8	1.02×10^8	35,200,000	8,320,000	293,000	Room air

ISO classes are governed by ISO 14644-1 while the Fed STD 209E is theoretically no longer valid but still very much in use.

weighed against higher capital equipment costs and stringent sterilization and operating procedures in the aseptic process. Cold aseptic filling also requires highly trained operators and this can become an issue depending on where the factory is located.

Blow Molds for Hot-Fill Bottles

Due to the thermal load on the blow molds they have been traditionally manufactured in stainless steel. Stainless steel is harder to machine, more costly, and has a lower heat transfer rate than aluminum, which makes it less suitable for standard blow molds. But for this application it was the material of choice for the longest time as it supports the high temperatures whereas the lifespan of most aluminum molds may be very limited. Newer developments have started to replace it with aluminum (see next subchapter).

The heating channel layout in the blow mold is much more elaborate than in a standard mold. Because of the low heat transfer rate of stainless steel cooling channels must be very close to the cavity walls, allowing rapid heating of the PET inside the mold. Mold designers must decide on the blow mold temperature beforehand and make allowance for their shrinkage calculations, as the mold will expand when heated. Stainless steel's lower thermal expansion rate helps minimize this effect. Because of high temperature and pressure all channels and sealing surfaces must be precision machined (Fig. 9.20).

Hot-fill bottles require extensive venting. The soft PET flows much quicker and easier, making air entrapment more likely. Hole vents of approximately 1 mm (0.040 in.) diameter are used in threads, on vacuum panels, and bases of the bottles. These days they are often produced by

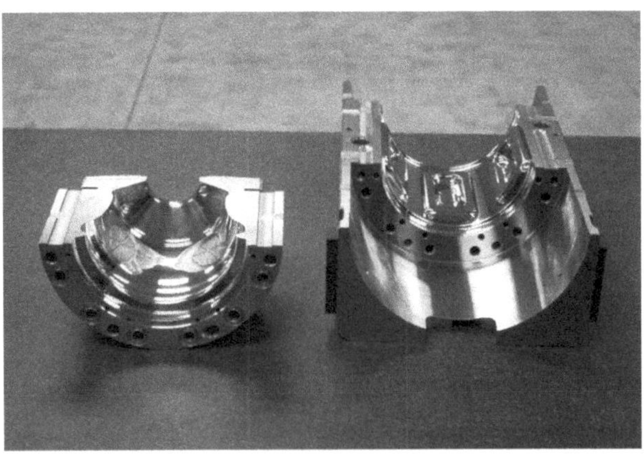

Figure 9.20 Cooling channel layout in molds for hot-fill bottles is complex and requires precision machining. Photo courtesy of Garrtech Inc.

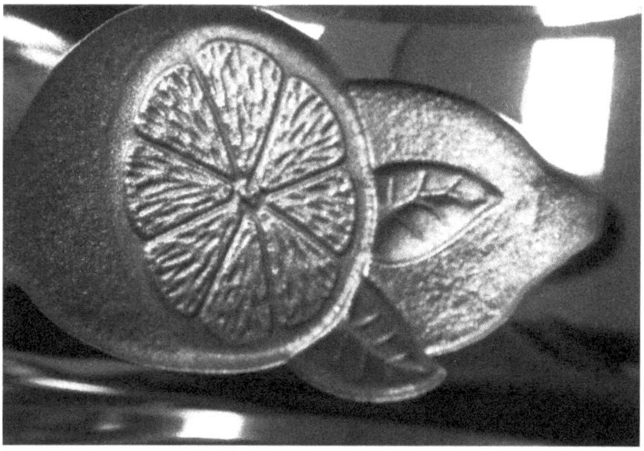

Figure 9.21 Higher preform temperatures allow the incorporation of fine detail work on mold surfaces. Photo courtesy of Garrtech Inc.

electrical discharge machining rather than drilled as the more economical manufacturing method.

The prolonged use of low viscosity oil to heat the blow molds to high temperatures leads to a breakdown of the oil with carbon deposits over time, narrowing channel diameters and reducing their effectiveness. Blow molds must be thoroughly cleaned at intervals specific to the temperature to which they are exposed and some companies send them back to the mold maker for this purpose.

One interesting aspect of the hot-fill process is that the softer material flows into fine detail much more easily than PET at regular blowing temperatures. This allows stunning detailed relief to be incorporated into bottle design, adding to its attractiveness (Fig. 9.21).

Blow Mold Coating

Cavity surfaces must frequently be cleaned of resin residue and aluminum typically does not have sufficient durability. However, newer developments have found ways to use high-grade aluminum and coat it with a variety of materials. Several methods exist to extend the life of aluminum molds but can also be used for stainless steel ones (Fig. 9.22).

- Nickel–Teflon
- Ceramic–nickel
- Diamond-like carbon coating (DLC) applied as
  - Physical vapor deposition (PVD) or
  - Plasma-assisted chemical vapor deposition (PACVD)

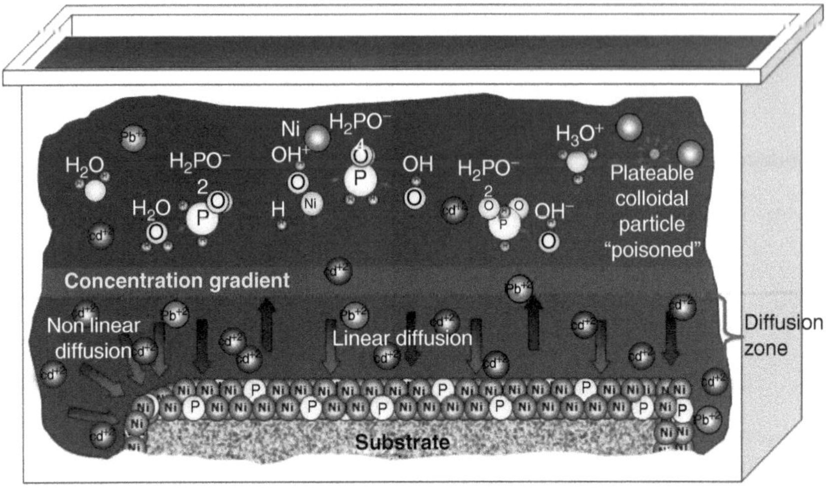

Figure 9.22 The electroless nickel (EN) plating process autocatalytically deposits nickel alloyed with phosphorous. There is no current resulting in uniform coating thickness. It can be enhanced with nanoparticles (ceramic, diamonds, PTFE, silicon carbide) for special deposit characteristics. Diagram courtesy of Compound Metal Coatings.

Nickel–Teflon (codeposition of electroless nickel and PTFE) is probably the most commonly known method.

Its characteristics include the following:

- 33 RCH as plated;
- 54 RCH postheat treatment;
- deposit density of 6.3–7.1 g/cm^3
- nonmagnetic property
- coefficient of friction: 0.1

Due to its low friction factor it improves release and antistick properties.

Ceramic–nickel is a less known coating with some interesting differences to nickel–Teflon. Its characteristics include:

- lowest coefficient of friction (COF) of all composites tested (decreases with increased load);
- uniform particle distribution;
- unlike PTFE (Teflon), the Cer–Ni coating is very hard and resistant to temperatures up to 3000°C;
- excellent release characteristics;
- very easy to clean; and
- selflubricating.

Both *vapor deposition* methods are 2–3 times more expensive than the other methods. Their characteristics are:

- wear and abrasion resistance;
- low friction (0.05–0.1);
- high hardness;
- excellent corrosion resistance;
- gas barrier;
- precision control of thickness;
- "line of sight" process: Part rotation is required and does not coat internal diameters.

This latter point is a disadvantage of these methods. They cannot coat the inside water channels and prevent corrosion and mineral deposits.

Materials for Hot-Fill Bottles

Several manufacturers have responded to the increased market share of these bottles and produce materials uniquely suited to the application. Some of these materials have a high intrinsic viscosity (IV) of 0.87–0.89 dL/g. There will be some IV degradation during the process and the high initial IV guarantees a sufficient final IV in the product. Other materials stay with the more common IV of about 0.82. Some have incorporated UV barriers of up to 390 nm into their resin. AA levels are less critical for these resins because of the sugar content of the filled product and it is often cited as 2 ppm. This is in contrast to low IV resins for water bottles that show AA levels of less than 1 ppm.

Another common feature of these resins is a higher glass transition temperature (TG). The TG of a material is that temperature range where the material loses many of its properties and generally becomes soft. For PET the value most often cited is 76°C (169°F). Filling temperatures around and above this level can distort the neck of the bottle making the abovementioned neck crystallization necessary. Resins with a TG of 83°C are commercially available and can often avoid this procedure.

These resins also show increased thermal stability leading to reduced shrinkage. This can also translate into material savings and a more stable bottle fill level. Materials differ in whether the preform temperature or the blow mold temperature has more impact on enhanced bottle properties and processors need to carefully study material specifications to make optimum process adjustments. Converters buying preforms for this process should demand to know what material was used in their manufacture so they can make the necessary process adjustments.

Panel-Less Designs

Brand owners have long complained about the appearance restrictions forced on them because of the vacuum panels. Labels crinkle when consumers grab the bottles as the panels are recessed and the labels only adhere to the ribs in between. There are now two designs in the marketplace that do not require traditional vacuum panels. Instead, both use the base of the bottle to account for shrinkage alone. Traditional heat-set bases are rather thick and feature ribs to avoid distortions. The panel-less designs aim to control base distortion thus eliminating the need for any other measure of vacuum replacement.

The Powerflex bottle features rigid sidewalls with a thin base. As a result of this wall thickness distribution the base pulls up when vacuum sets in allowing the sidewalls to stay straight. The ratio of sidewall to base thickness must be kept around 2.5:1, which translates in sidewalls being around 0.63 mm (0.025 in.) thick and bases around 0.25 mm (0.010 in.). As there is a limit on how thin a base can be blown and still withstand considerable pressures, these bottles weigh in at the top of range of hot-fill bottles making them more costly (Fig. 9.23).

Another way of getting around vacuum panels is adding a gas to the headspace whose expansion is meant to compensate for the negative pressure inside the bottle. Using special microdosing units it is possible to add liquid nitrogen to the headspace just after filling. The nitrogen replaces the oxygen in the headspace, warms up, and fills the volume that is lost during shrinkage. It thereby increases the pressure inside the bottle just after filling and may so reduce or equalize the vacuum acting on the bottle walls. However, if the dosage is too high or the filling temperature increases during a line stoppage the bottle may also balloon out as a result and not shrink back during cooling. This technology holds the promise of reducing container weight, eliminating vacuum panels, and so opening up new opportunities for hot-fillable PET bottles.

9.5 Preferential Heating

One of the great advantages of the stretch blow process is that all preforms enter the machine with the same uniform temperature and the oven system in the blow machine then reheats these preforms to an optimal temperature profile that is uniform around the circumference of the preforms in any one spot. But what when this uniformity is not ideal for the blowing process? There are many bottles that are oblong, that is, they are oval, rectangular, or irregular in some other way.

Figure 9.23 This hot-fill bottle uses the bottle bottom instead of vacuum panels to counter the effects of vacuum. Photo courtesy of Amcor Rigid Plastics.

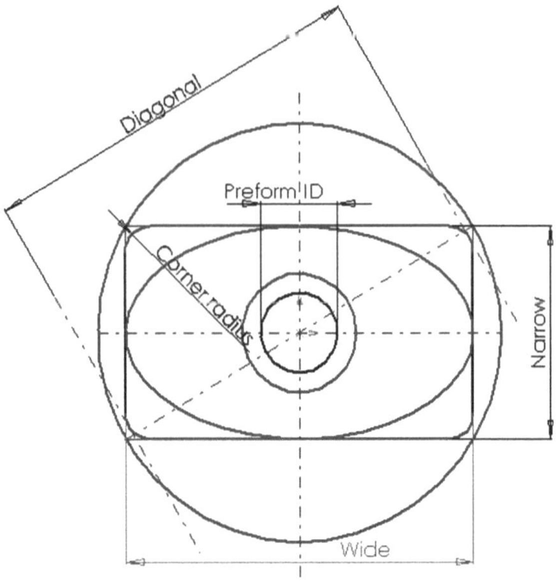

Figure 9.24 Irregular shaped bottles require different calculation methods of the hoop extension.

When the aspect ratio (i.e., wide container dimension divided by the narrow dimension) is greater than 2, preferential heating may be called for and will improve container wall thickness distribution significantly. At lower aspect ratios preferential heating will have a lesser impact albeit always a positive one (Fig. 9.24).

When preforms are reheated equally around the circumference all parts of the preforms inflate equally as well. As the container walls of an irregular container are of different distances from the center, the inflating preform reaches the walls nearer to the center earlier than those further away. This leads to different stretch ratios with the foreseeable result that material hitting the nearer walls stays thicker than material traveling a longer distance and thus stretching more. One can often see and feel a so-called spine in the center of the wide panel of an oval bottle, that is,. a vertical, thick area that then shrinks more than the rest of the bottle leading to problems with printing or labeling (Fig. 9.25).

The solution to this problem is to reheat preforms unevenly in a way that heats the areas of lesser stretch more than areas that have to stretch further. In a way, the cooler areas then pull the warmer ones along so leading to a more even wall thickness distribution. There are several ways of imparting preforms with a preferential heating profile. Depending on machine type and manufacturer two solutions have emerged. One is to heat

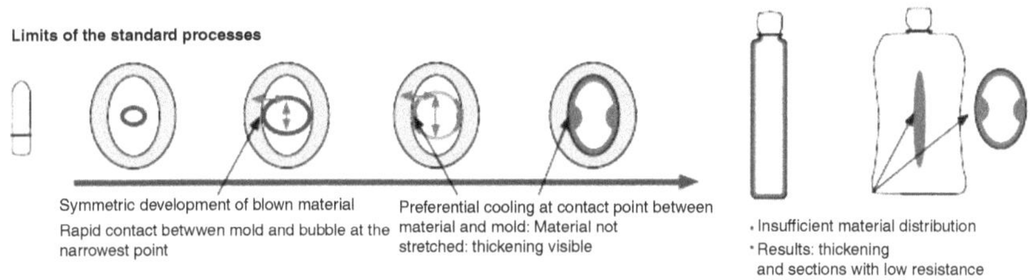

Figure 9.25 Limits to wall thickness distribution of the standard process. Diagram courtesy of KHS Corporplast.

the preform areas that will form the short sides of the container more than the ones forming the wide ones. The other solution is the opposite: cooling the preform areas that form the wide container sides. From a process standpoint the latter method has some merit as the outside of the preform is cooled and as we have seen in Chapter 7, Section 7.2 the optimal heat profile in the preform wall features a higher inner temperature (Fig. 9.26).

The most common method on a rotary blow machine is to spin preforms as usual for part of the way but at some point the mandrels are rotating on are locked and stop spinning. Preforms then travel a certain distance under these conditions and are unevenly heated. This orientation has to be extended into the blow mold so that the hotter preform areas are in the right spot when the mold is closed. This can be done in a variety of ways, each unique to a particular manufacturer (Fig. 9.27).

The second method uses special reflectors that feature reflective and nonreflective areas in the same pattern throughout part of the oven system. This assures that those preform areas that continuously pass by the nonreflective areas receive less heat and are then cooler when they enter the blow mold. This orientation has to be kept with special grippers. Standard grippers move toward the preforms and open up on spring-loaded mechanisms

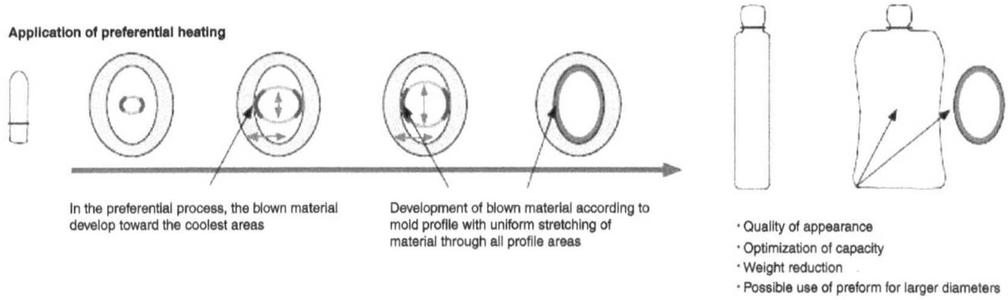

Figure 9.26 Preferential heating improves wall thickness distribution. Diagram courtesy of KHS Corporplast.

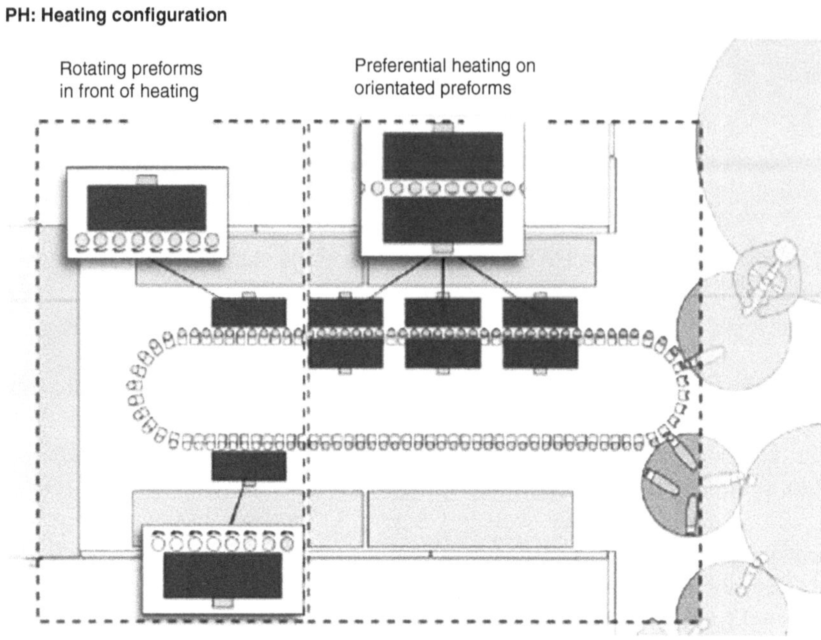

Figure 9.27 The most common configuration is to prevent the mandrels from spinning and heat them from both sides. Diagram courtesy of KHS Corporplast.

while they grab them. This could lead to a turning of the preform which is not permissible when they are oriented in a way that the warmer and cooler areas match the respective areas in the bottle mold.

The advantage of the former method is that a machine equipped with a preferential oven system can also produce round bottles without any change in the machine setup whereas the latter method requires replacement of the special reflectors with standard ones.

Chapter 8, Section 8.5 shows how this is accomplished in the single-stage process.

9.6 Direct Feeding of Preforms into the Blow Machine

General Observations

In the standard two-stage process preforms are injection molded on completely separated machines that may be in the same building or on different continents. They are completely cooled down before they go into the blow machine which assures two things:

- There is no temperature difference between any preforms.
- There is no temperature difference within any preform.

Typical preform temperatures when they exit the injection machine (after postmold cooling) are between 45 and 60°C (113–140°F). There is no obstacle of them being used at that temperature at the entry to the blow machine as long as the two conditions above are met because the blow process can be adjusted to suit. However, meeting these at those temperatures is difficult because of the effects of viscous heating and the minor imbalances in the cooling water flow of the injection molds.

There is a considerable amount of handling, packing, and shipping involved in the standard model. This leads to labor and transportation costs that could be saved if preforms were fed directly. Additionally, since many preforms look very similar there is always a chance of mistaken one for the other with the subsequent interruption of operations.

On the other side the standard model has proven very flexible. Many brand owners that use in-house production and a sizable proportion of converters have opted to stay out of preform production, which requires more technical talent and has a steep learning curve whereas stretch blow molding is a somewhat easier trade to learn. It is often difficult to match outputs of injection and blow machines and separating them allows companies to run one when the other is down.

In order to successfully implement a direct-feed solution, the following conditions should be met:

- Output of the injection machine should exceed that of the blow machine by 5–10% or more.
- All overflow preforms must be collected and cooled down in the standard way. These can be used in the same blow machine at a time when the injection machine is down (albeit with a different set of process parameters) or in another machine if that is an option.
- A limited number of tools. Due to complex setup no more than three tools should be used preferably with the same neck configuration to reduce setup times. Very effective is a situation where more than one bottle is made with the same preform.
- Preform wall thickness should be below 3.5 mm to avoid overly long conveyors. If injection and blow machines are physically apart a great distance (over 60 m) then this is less of an issue.

Preform Temperatures

When preforms leave the injection machine they have three different temperatures: inside, outside, and center. Infrared sensors and thermocameras can only measure the skin temperature. We therefore have to rely on prolonged observation and mathematical models to come to an understanding of the other two temperatures.

Let us first look at the way the melt is cooled down in the tool. After the material has completely filled the cavity hold pressure pushes the molten resin evenly against the core and the cavity. After hold time (during the cooling time) this pressure is relieved and the resin naturally shrinks onto the core. This makes core cooling more effective even though cavity cooling still contributes. From these facts one could conclude that the inside temperature is somewhat lower than the outside temperature when the preforms are ejected. However, this is not the case. I have personally measured these temperatures by immediately cutting preforms in half after ejection and measuring the surface temperature of both inside and outside. The inside always came in at a higher temperature, by as much a 8°C. (This explains some of the great containers one can make in the single-stage process.) The reason for this discrepancy is twofold:

- The surface area of the inside preform is always smaller than that of the outside. The thicker the preform wall the more pronounced this effect becomes. Lower surface area means less cooling area (Figs. 9.28 and 9.29).
- There is usually more room for cooling water channels in the cavity than there is in the core. For a typical 28-mm neck with an inside diameter of 25 mm, the inside diameter of the core cannot exceed 19 mm leaving 3-mm steel wall thickness. Water enters the core through a so-called bubbler tube that takes another 1.5 mm away from the total. A balanced in and out flow may lead to this configuration.

  Both inflow and outflow area of this configuration are around 125 mm^2. Note the cross section of the outflow area at 2.25 mm. In cavity cooling this can be increased to 3 mm contributing to more cooling water flow. Combined with the larger surface area this leads to much more effective cavity cooling.

There is another complication when assessing preform temperatures: the center of the preform cools down last as it is cooled from both inside

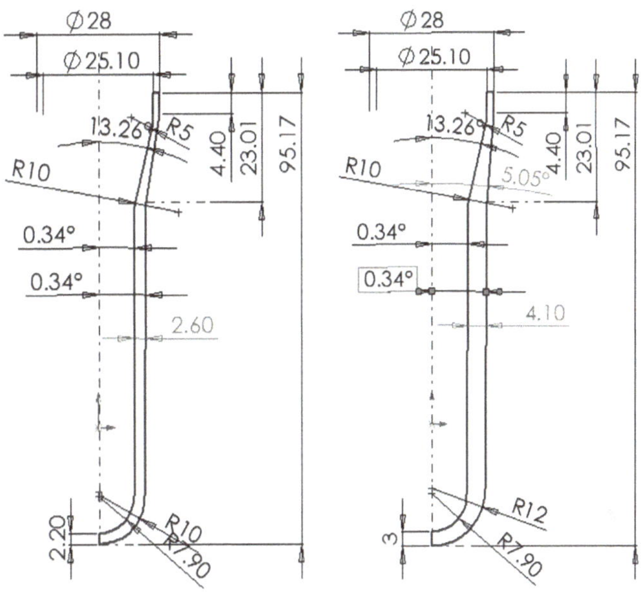

Figure 9.28 The inside surface area of both preforms (shown without neck) is 5.6 cm². The outside area of the left preform is 7.3 cm², while that of the right preform is 8 cm². This equates to a reduction of inside surface area of 23 and 30%, respectively when compared with the outside area.

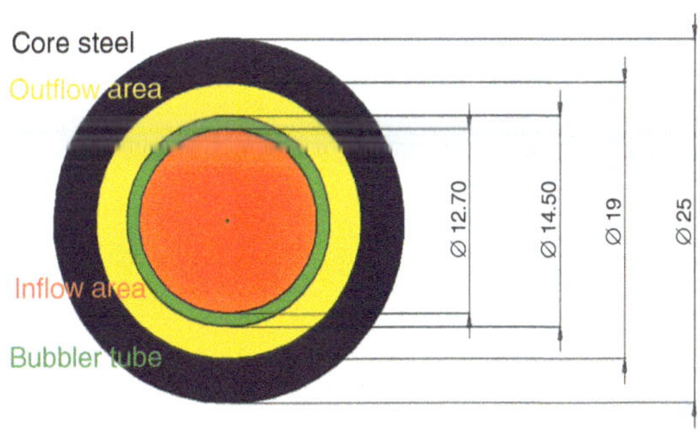

Figure 9.29 Cross-sections of assembled injection cores.

and outside. Because of more efficient cavity cooling the hottest part of the preform after ejection will be closer to the inside than the outside. How much warmer this part will be depends on processing conditions and especially wall thickness. Preforms are generally ejected as soon as possible in order to reduce cycle times. For successful ejection the skin temperature must be below ~75°C. If it is higher the preform may stick in the cavity or on the core both situations leading to deformed or damaged and in both cases unusable preforms. Thicker preforms retain more residual heat and therefore need longer postmold cooling times.

If all preforms cooled down in the exact same way, this would not concern us too much. Unfortunately, there are significant temperature differences between preforms of the same injection shot. Data of an actual application that was investigated is mentioned here. Preforms were free dropped, that is, without postmold cooling and infrared pictures taken about 1 min after. Process data were as follows:

Cycle time	20.7 s
Injection time	1.8 s
Hold time	2.2 s
Cooling time	12 s
Wall thickness	3.2 mm

The infrared image shows a number of preforms 1 min after ejection (without postmold cooling) (Fig. 9.30). The scale has been chosen so that the necks do not show variations between 73.9 and 63.3°C are visible. If fed to the blow molder at this stage, large bottle wall thickness variations would result.

After 24 min the temperature differences are now less than 1°C while preforms are on average still 3°C above the room temperature (Fig. 9.31).

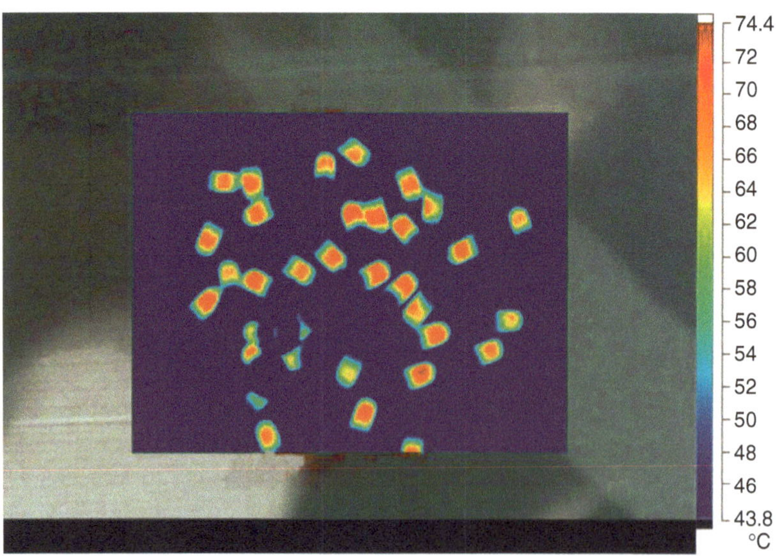

Figure 9.30 Ejected preforms show considerable temperature differences.

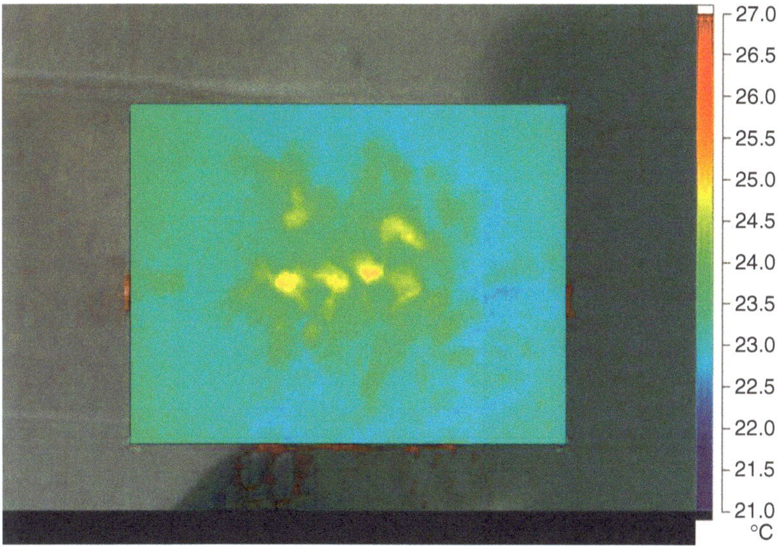

Figure 9.31 Temperature differences have evened out after 24 min.

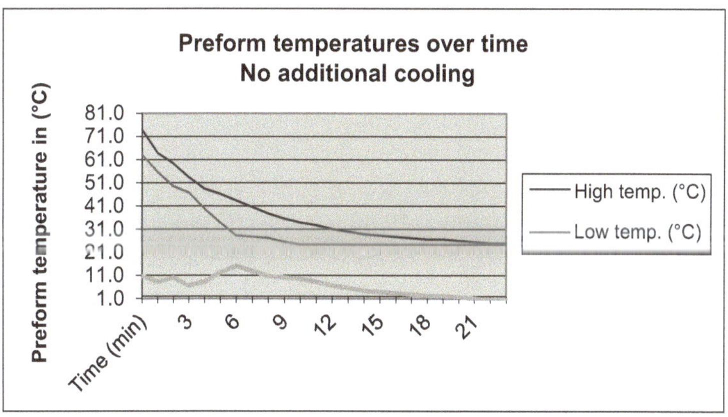

Figure 9.32 Preform temperature development without cooling.

Fig. 9.32 shows how the low temperature becomes stable after about 10 min whereas the high temperature takes another 14 min to come within 1°C. It is not apparent if the differences stem from within preforms or are differences between preforms. For this application it makes no difference but would be important to know if the intention was to reduce those (Fig. 9.33).

By exposing the preforms to fan cooling the time to cool down to a temperature difference of less than 1°C is shortened to 14 min.

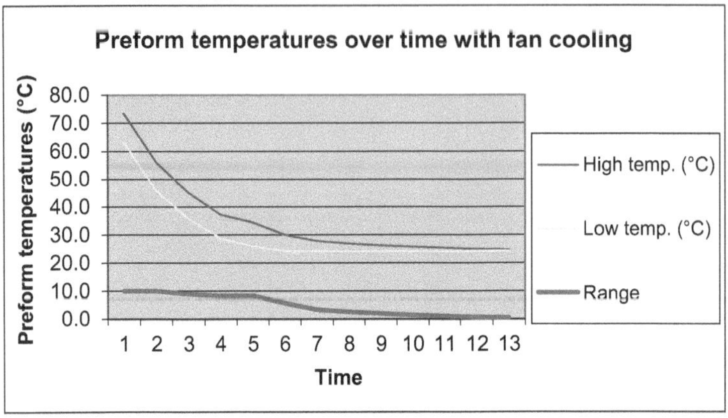

Figure 9.33 Preform temperature development with cooling.

Conveyors

It is clear from the above data that conveyors with top-mounted cooling fans will reduce the necessary cooling time significantly. As mentioned previously the collected data is highly dependent on preform wall thickness; therefore the preform with the thickest wall will determine the required conveyor length if this is not a function of the physical position of the injection and blow machine alone. To improve cooling performance of the fans unchilled water condensation circuits may be added that can effectively and at little operational cost cool fan air down by up to 15°C (depending on the temperature difference between plant air and water temperature).

If this is not an option and the machines are situated too close to each other to directly connect them, a circuitous conveyor track underneath the ceiling can be designed to give the desired length. There are two parameters that determine the conveyor length after the residence time has been calculated:

- the width of the conveyor and
- the speed of the conveyor.

The most voluminous preform in the highest cavitation will determine how much room a full shot of preforms will occupy. It may be neck size or length that is the most crucial feature. Trials should be undertaken to make sure that there is room for the preforms to lie flat without being too crowded so as to facilitate cooling. The optimal conveyor speed is when the conveyor is fully loaded with no empty spaces on it. This in turn depends on the room a full injection shot will take and the cycle time of the injection machine. Here is an example:

A 96-cavity machine runs at a cycle time of 11.5 s. This leads to an output of 500 preforms/min. A full shot requires 965 mm (38 in.) of conveyor length. The design speed of the conveyor is therefore:

500 preforms/min/96 preforms/shot × 965 mm/shot = 5026 mm/min or 5 m/min (16 in./min)

This value is then multiplied by the number of minutes of residence time that has been previously determined. Residence time of 20 min would require a conveyor length of about 100 m (305 in.).

Layout

While every installation will be somewhat different depending on the individual plant characteristics there are some common features. After the initial drop-off preforms will be conveyed upward and out of the way. The cheapest solution for this task are cleated conveyors like the ones used to carry preforms from the blow machine hopper to the unscrambler. If there are crane installations around the machines great care must be taken not to cause any interference. They will then travel a certain distance underneath the ceiling, maybe traverse a wall to the separate blow molding room and eventually end up above the blow machine hopper. There needs to be a sensor to indicate a "hopper full" condition. When this sensor is active preforms need to travel to a packaging crate (usually a cardboard box of roughly 1 m^3 volume, also called a gaylord) from where they are manually removed. (Of course this could be automated as well.) The easiest way to do this is have preforms dump onto a short final conveyor whose direction can be reversed so that at one end preforms fall into the hopper while they fall into the crate at the other, all controlled by the "hopper full" sensor.

Needless to say that preforms collected in the crate and allowed to cool down completely cannot be used in the same blow machine setup as the directly fed ones. As an added safety measure an infrared sensor can be mounted to the in-feed rail of the blow machine. This sensor continuously measures the skin temperature of the preforms that are about to be blown. Once the process has been established the output of this sensor can be connected to either a standalone or the machine controller and stop the machine from feeding when the preform temperature is out of range. To automate this process further a gate can be constructed in such a way that the out-of-spec preforms are diverted into another crate until the preform comes back to the right temperature. An algorithm with a moving average will even out small temperature differences. In any case, whenever the sensor detects a fault condition, the machine should alarm the operator.

A typical scenario for this kind of fault would be a malfunctioning of the injection machine cooling circuit that is too small to be noticed on the injection end but big enough to cause a disturbance on the blow machine. As processors understand the process for a particular preform better over time and know which parameter change leads to faulty bottles they usually make corrections to these kinds of parameters.

Operation

Before continuous feeding can begin, the blow process should be optimized for standard cold preforms. This setup can be saved and used whenever overflow preforms are fed into the machine and it will reduce the time it takes to establish the setup for the direct-fed preforms. This setup will use less heat as the preforms are still warmer in the center of the wall. The trials can be done in the same way as usual where a set of 4–20 preforms is tested before the feed is continuously energized. This will lead to inconsistencies as preforms will spend different times in the hopper and the processor must take this into account. If the injection produces slightly more output than the blow machine can process (the optimal configuration) the hopper will eventually be full and nearly full at all times giving preforms additional cooling time. However, it is paramount that all processing changes to the injection machine are communicated to the blow machine operator(s) so that necessary changes in the blow machine setup can be made.

In a standard blow machine there are sensors on the preform in-feed rail that eventually shut feed to the blow machine off when preforms are not present. The machine then keeps turning but reduces lamp output to a predetermined value while alarming the operator to a problem on the in-feed line with a visual and/or auditory signal. To allow fully automatic operation the blow machine has to be modified by adding a "direct-feed" button or similar. If this function is selected the blow machine must restart itself when the sensors show preform presence again. A lack of preforms may occur when injection machine operators do routine maintenance or have a temporary machine breakdown. To distinguish this situation from a fault of preforms stuck in the unscrambler or on the in-feed rail (something that occurs far too often even in modern machines) a second sensor is needed that indicates a "hopper empty" situation. The logic should then go as follows whenever a lack of preforms is detected and the "direct feed" button is selected.

- If the "rail empty" and "hopper empty" switches are both made the machine should reduce lamp output but keep running without any alarm indication, then restart when preforms become once again available.

- If the "rail empty' sensor is made but the "hopper empty" sensor is not the machine should alarm the operator to a preform jam.
- The "hopper full" sensor only controls the direction of the last preform conveyor before the blow machine and must be connected to a timer so as to avoid constant switching.

Economical Considerations

The question of how much more economical a direct-feed solution can be must be addressed in the initial planning stage. Some of the issues are subtle and require careful investigation. Foremost are labor savings. A well-designed system requires no labor between resin in-feed to the injection machine and bottle out-feed at the blow machine. Of course, this system can be further enhanced with automated vision systems to inspect bottles and palletizing equipment. Vertically integrated fillers have added filling, labeling, packing, and palletizing to the downstream equipment and we will look at this in the next paragraph. For now we will just investigate where all the labor savings can be found in an injection to blow system.

In a typical medium-sized plant preforms fall into the aforementioned gaylords and are manually removed to a storage area. Depending on demand requirements they are in storage for a few hours or several days/weeks and are totally cooled down when they are removed and dumped into the hopper of the blow machine. There is actually little labor involved, most of the costs are comprised of storage containers and space. Going back to our 96-cavity system let us assume it produces a 23 g preform with a 28-mm neck and that 15,000 of those fit into a gaylord. At 500 preforms/min, a forklift has to be dispatched every 30 min. It will be a 5-min run and let us say another 10 min to bring the preforms to the blow machine and operate the dumper. So that is 15 min saved for every half hour of production or about \$7 in wages/h if we assume an hourly wage with benefits of \$14.

Next comes storage space. Assuming a storage time of 24 h and storing three gaylords on top of each other. The 48 gaylords filled with 1 day's production will take about 16 m^2 (176 $ft.^2$) of space. At a monthly rent of \$5.50 m^{-2} (\$0.50 $ft.^{-2}$) this will translate into \$0.10 h^{-1}.

There is some initial investment for the pallets and gaylords, about \$12 per set. Monthly losses may be in the 25% range. A company may have to buy 100 of these sets losing 25 per month. Neglecting the initial investment a monthly cost saving of \$300 can be attributed (\$0.42 h^{-1}). Add to that the elimination of the polyethylene liners @ \$0.80 per piece for a total saving of \$1.60 h^{-1}. This gives us hourly savings of:

Labor	\$7
Storage	\$0.10
Gaylords/pallets	\$0.42
Liners	\$1.60
Total	\$9.12

In a 24/7-type production, this will lead to about \$72,000 in yearly savings, about equivalent to the cost of implementing this system. In short, the ROI is about 1 year but can be significantly more if large conveyor distances are required.

Other advantages may be

- Elimination of forklifts and drivers.
- Reduced dust from cardboard elimination.
- Less operational issues from overflowing gaylords or incorrect preform allocation.
- Slight improvement in aesthetic qualities as preforms are much colder when they fall for the first time a great distance and they fall only once whereas preforms made with the standard model fall twice: first into the gaylord at elevated temperatures and then again in the machine hopper.
- Possible elimination of other line personnel that is required to make sure gaylords do not overflow.

Other Automation

Once a direct-feed system is considered why not go a step further and add vision inspection and palletization to the system. Some fillers have added all downstream equipment to an installation with great success. Here are a few considerations:

- Vision inspection of bottles and possibly preforms is paramount to get 100% inspection and eliminate out-of-spec bottles getting into the hands of customers.
- Only very reliable machines should be selected. If one fails a whole line is down at a higher loss of revenue.
- Downstream equipment should be 5% faster than upstream equipment (except the blow molder as discussed earlier).

- Downstream equipment controls upstream machinery (except the injection machine). When there is a fault condition at the filler, the filler will stop the in-feed to the blow machine and turn it back on once it is operating normally again.
- Storage silos or buffers for bottles are needed to allow the blow machine to empty preforms inside the machine in case the downstream equipment signals a "stop" command. Depending on the cavitation several hundred preforms are inside a blow machine and would have to be thrown away if there was no room for them as bottles on the line to the filler. The slightly higher speed of the filler (or the palletizer) will allow bottles in the storage area to be worked off and make room for new bottles.

Conclusions

Direct feeding of preforms does not suit every company and every application. Changing these lines over twice a week may pose too much of a challenge to most companies. Running all year round with the same preform and maybe a few different bottle shapes is ideal and can lead to better quality and lower-cost bottle production.

9.7 Vision Inspection

Quality control has become increasingly important since outputs of blow molders are now routinely in the 30,000 bottles/h range with machines capable of producing 72,000 b/h as of this writing. The sheer amount of noncompliant bottles over even a short period of time can cause severe problems in the manufacturing plant. Imagine, for example, that a preblow valve malfunctions intermittently (a not unusual behavior) and all bottles from this one cavity exhibit neck folds as a result. QC intervals might be 2 h, and it might take another 15 min before the defect is discovered and the machine stopped. After 2.25 h there are now 67,500 bottles that must either be thrown out or manually checked, a very time-consuming and expensive undertaking. The chance that an operator will find this defect in a timely manner are very slim whereas the automated system will not only discard all bottles that are out of spec but also alert the operator to problems in the specific cavity where the problem is happening. Vision inspection systems derive their value from these scenarios. They will discard faulty bottles, alert operators to a problem, and even point out the exact cavities with high

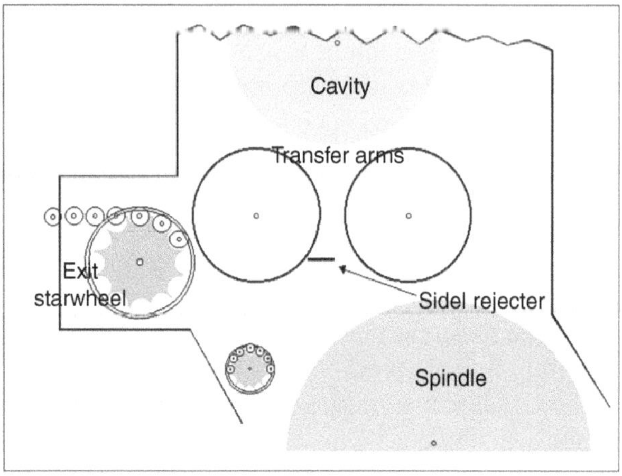

Figure 9.34 Vision inspection installed inside a Sidel blow molder. Diagram courtesy of Pressco Technology, Inc.

reject rates. All of this happens in real time and every bottle is checked. The often-postulated zero-defect production can now become a reality.

System Overview

All systems use some type of camera, a light source, and/or an emitter/receiver module. Cameras can be mounted above or beside the passing bottles depending on the area being monitored. The analog resolution of the camera is digitized into a finite amount of discrete pixels with the help of a frame grabber. Proprietary software processes the resulting digital image, reduces noise, and evaluates it for possible faults.

Vision inspection can be online, mounted directly inside the blow molder or offline, mounted alongside the conveyor belts on which the blown bottles travel. Installed inside the machine, the systems use the blow molder's own reject system to discard bottles. It takes approximately 33 ms to complete the imaging process, leaving enough time for the reject system to be engaged as even at 72,000 b/h there is one bottle every 50 ms (Fig. 9.34).

When installed externally, a single bottle stream may be diverted into two streams if the machine produces more bottles than the inspection system can handle. External systems use their own rejection systems and require some floor space, which can be seen as a disadvantage over systems installed inside the machine. However, they can be adjusted or repaired if necessary without stopping the blow molding machine, assuming that plant personnel take the risk of running without inspection (Fig. 9.35).

Figure 9.35 Vision systems may be equipped with up to six cameras to check on all features of the bottles. Photo courtesy of Intravis Vision Systems.

There are also very comprehensive systems that can be used to measure "one shot," that is, a full set of bottles or preforms from a machine and measure them automatically. This is then done in lieu of having quality control personnel do it manually. As cavitations of both injection and blow machines steadily increase (maxima are 216 for injection and 40 for blow as of this writing), quality personnel cannot keep up with the sheer amount of parts and have resorted to measure only subsets of cavities. While this makes sense from an operational point of view that before-mentioned faulty preblow can so affect a much larger number of bottles and cost the molder some serious money (Fig. 9.36).

Applications

Inspection systems monitor a variety of bottle parameters. Areas that can be monitored are:

- neck sealing surface and ovality;
- sidewall defects such as folds, holes, etc.;
- base defects such as cracked or off-center gates, holes, etc.;
- contamination anywhere in the bottle; and
- sidewall thickness (Fig. 9.37).

Monitoring neck ovality before the preforms enter the blow machine is extremely important for high-speed blow molding machines. Preform necks may be crushed during storage and transport, and become deformed.

Figure 9.36 Vision inspection installed on two separate bottle streams outside a blow molder. Picture courtesy of AGR International.

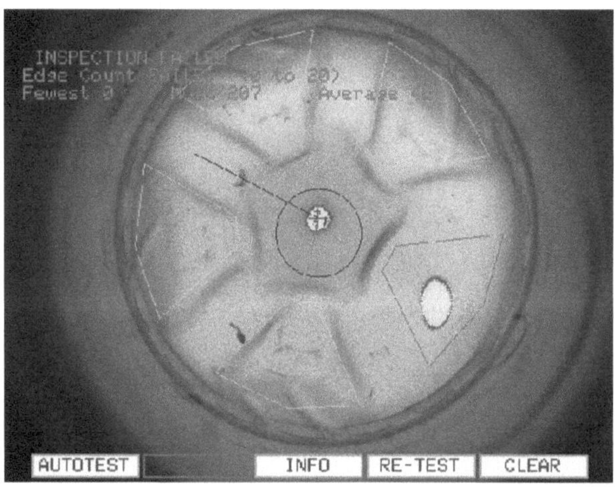

Figure 9.37 Screen shot showing base with hole. Photo courtesy of Pressco Technology, Inc.

If the deformation is severe enough the preform will cause the in-feed to jam or it may break during placement on the mandrel and later fall off and cause a jam that way. In either case production must be stopped, the jam cleared, and hundreds of preforms that are in all sections of the machine must be discarded. Inspection systems check each preform neck before it is ever placed on the mandrel and defective preforms are effectively

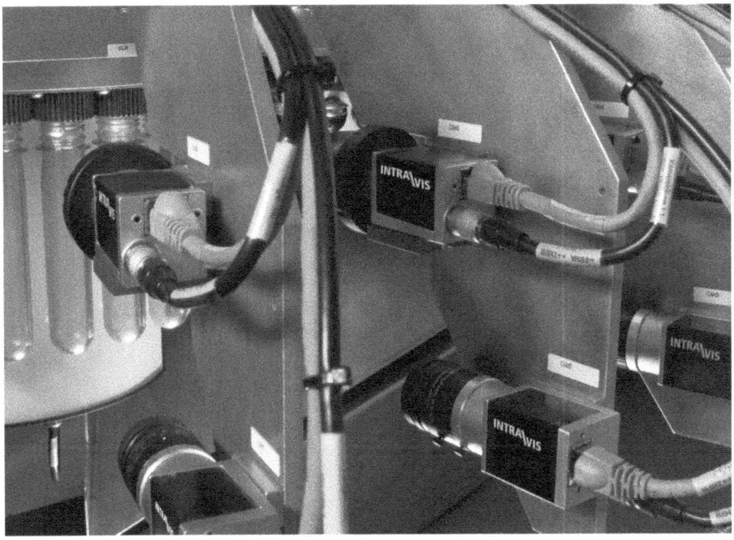

Figure 9.38 Preforms may also be inspected either at regular intervals at the production plant or in spot checks in the blow molding plant. Photo courtesy of Intravis Vision Systems.

disposed of. Again this system may be placed right inside the machine or on the in-feed chute (Fig. 9.38).

Beyond Vision Inspection

In a typical plant operators or QC personnel will take samples at routine intervals and check wall thickness in several locations besides other examinations. Processors use this information to adjust the heat profile and/or blow parameters if the wall thickness does not meet the specifications. This intervention requires a lot of knowledge and can be a time-consuming exercise. There are two steps involved and there are now solutions for both. The first step is to automatically record wall thickness data eliminating the need for manual measurements.

This is done with infrared modules that are placed inside the blow machine. First models only measured two or three locations (depending on the height of the bottle). The newest system uses a module that can measure up to 32 wall thickness points on the bottle (again depending on the bottle height) giving plant personnel a multitude of data points to fine-tune bottle wall thickness in as many spots. These measurements are accurate to 0.02 mm (0.0008 in) and repeatable to 0.003 mm (0.0001 in.). Each spot has its own target and upper and lower limit and the operator is

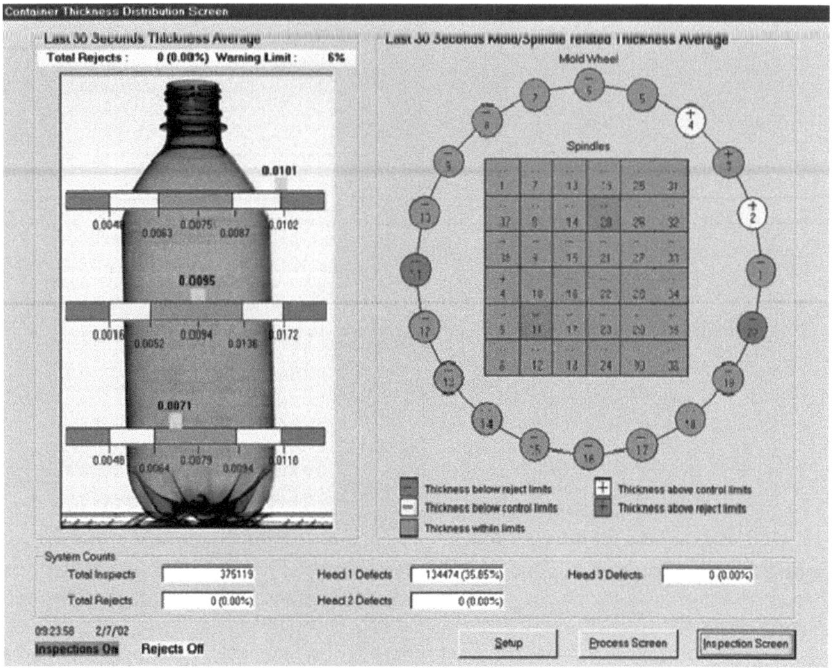

Figure 9.39 This system uses up to 32 infrared sensors to measure wall thickness of blown bottles. Picture courtesy of AGR International Inc.

alerted whenever one spot moves out of the set zones. Statistical data such as *C* and *R* plots are also available allowing operators to make changes whenever a trend toward noncompliance is detected (Fig. 9.39).

One caveat should be mentioned, the infrared sensor looks through both bottle walls dividing the measured thickness by two and giving the operator this value. However, as explained in a later section, if the preform gate is not centered correctly, one side of the bottle may be significantly thicker than the other. The infrared sensor may show this situation as correct if the total value divided by two is within the selected limits. This problem can only be detected with a camera that looks from the top or bottom into the bottle and detects the off-center gate.

Another approach is taken by actually measuring the weight of the base in real time. This is a novel idea taken from the widespread practise of section-cutting bottles to monitor performance of the system (Fig. 9.40).

The second step is to automate the necessary process changes whenever a deviation from the chosen wall thickness profile is detected. This requires access to the blow molding machine's settings for lamp output, preform temperature measurement, blow air timing, and blow pressure.

Figure 9.40 This module weighs the base weight of every bottle comparing it with a set value. Picture courtesy of Pressco Technology, Inc.

Sophisticated algorithms can activate more than one control at a time, a method that is strongly discouraged in training sessions as processors often lose track of what is happening. However, a machine algorithm is capable to anticipate outcomes more reliably and quickly and is therefore not hindered by human comprehension capacities. The sheer amount of data, whether base weight or container wall thickness used as input, is also much easier and faster processed by a computer than an operator that will measure only a few points on the bottle and draw conclusions from this limited data pool (Fig. 9.41).

Another important feature of automated systems is that they measure every bottle, not just some samples every few hours. Day/night changes in plant temperature, air pressure, and humidity as well as changes processors made during preform production can all influence the blow process

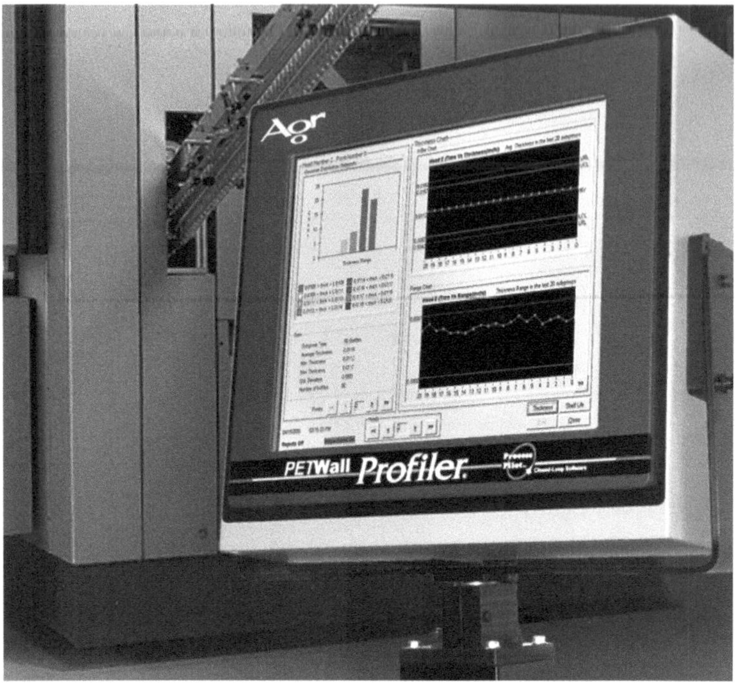

Figure 9.41 This system monitors bottle wall thickness and makes adjustments to blow machine parameters to keep it to programmed set points. Picture courtesy of AGR International Inc.

and with it bottle wall thickness in subtle ways that automated systems can cope with at least to some degree (Fig. 9.42).

Yet another approach is to have a camera look from the top of the bottle into the illuminated base and distinguish dark (heavy) and light (light) areas in the base and calculating these values to come with a base weight value. This information is then used to adjust a variety of blow machine parameters (Fig. 9.43).

There are a number of names that these systems are marketed under. Here are the ones mentioned in the text:

"PET Profiler" from AGR International Inc.

"Intellimass" from Pressco Technology, Inc, known also as "Equinox" by Sidel Inc.

"PETView" from Krones AG.

It is obvious that these systems will play a significant part in the future of blow molding. Since their price tag is virtually the same for any cavitation machine high-cavitation systems will offer the quickest payback and that is where the investment is going at this moment. This does not mean that plants

Figure 9.42 This system uses the base weight to make adjustments to the preblow settings. Picture courtesy of Sidel Inc.

will no longer need highly skilled processors. These rare specimens will still do the initial setup and decide how the material needs to be distributed in the bottle, or if the preform has problems, and will create the first machine recipes and be involved in new bottle development. But once a process has been established and verified the autopilot can take over reducing waste and down time and coming closer to the ultimate goal of zero defects production.

9.8 Barrier Enhancing Technologies

Why does PET need additional barrier enhancements?

It depends on the product that goes into the PET bottle and the shelf life it is required to sustain. All plastic bottles are to a smaller or larger degree subject to permeation of gases and moisture through the bottle walls.

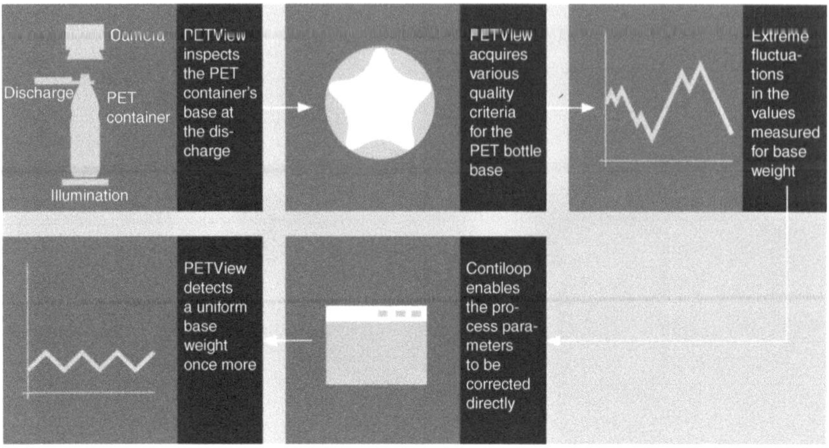

Figure 9.43 This system checks base structure from the top and makes adjustments to the machine to keep it in spec. Picture courtesy of Krones AG.

The basic background on matters of permeation is the spacing of the molecular chains. They are less than 1 nm (3.94×10^{-8} in.) apart whereas oxygen (O_2) and carbon dioxide (CO_2) molecules are in the 0.3–0.4 nm range. Over time, molecules like O_2 will permeate in (ingress) or permeate out (egress) as is the case with (CO_2) and so shorten the lifespan of the product inside. Moisture loss is another concern especially for products with "long" shelf lives (over 6 months).

Permeation Rate

The rate at which permeation happens depends on several factors:

- wall thickness
- orientation of the PET walls
- size of container
- temperature and humidity conditions of the filled bottles

It should be obvious that a thicker PET wall will reduce the permeation rate but this of course is not the preferred enhancing method in the age of lightweighing. What is less known is that highly oriented PET reduces permeation rate compared to amorphous PET (Table 9.4).

This situation has an impact on preform design. High stretch ratios add orientation and lead to better barrier performance and designers must take this into account (Chapter 6).

Table 9.4 Showing the Difference

Property	Unoriented (Amorphous)	Oriented	Improvement (%)
Thickness (mm)	0.36	0.36	n/a
Water vapor transmission rate ($g/m^2 \times 24$ h)	3.4	2.3	32
Oxygen permeability ($cm^3 \times mm/m^2 \times 24$ h $\times$ atm)	2.9	2.2	24
Carbon dioxide permeability ($cm^3 \times mm/m^2 \times 24$ h $\times$ atm)	15.7	14	11
Tensile modulus of elasticity (MPa)	3170	4960	56
Tensile stress @ yield (MPa)	82	172	110

The size of the container also has a large impact. To give an example, a standard 2-L CSD bottle features a shelf life of 6 months, whereas a 600 mL (20 oz) bottle is limited to 8 weeks under the same conditions. This is caused by the volume to surface ratio. The smaller the bottle the smaller this ratio. A proportionally larger surface area allows more contact between the product and the outside and therefore leads to more permeation, all other things being equal. Soft drinks brand owners resisted the 20 oz package for years because of the limited shelf life but eventually streamlined their distribution system to allow this package on the market. Today, it is the most successful bottle worldwide.

Storage conditions also impact shelf life. Higher temperatures lead to minute movements of the molecular PET chains and so facilitate permeation. High humidity acts as a lubricant allowing more CO_2 or O_2 molecules to slip by the PET chains. In Chapter 11, Section 11.10 there is a list of good manufacturing practices that outlines the dos and don'ts when it comes to storing and transporting PET bottles.

Product Types

There are a number of areas where the standard and already high (compared to other plastics) barrier performance may not be sufficient. These areas are as follows:

- carbonated soft drinks in small, <20 oz containers
- juices and functional drinks

- cosmetics
- other food stuffs such as ketchup
- milk
- beer
- chemicals

Carbonated Soft Drinks

Most CSD products are filled with a carbonation level of 3.8–4.2 by volume. Soft drink companies are concerned that the consumer experience is diminished when that level drops by 10–15%. This impacts the brand image and therefore must be avoided.

Juices and Functional Drinks

Brand owners in this market segment are mostly concerned with oxygen ingress, which leads to vitamin loss, and color and taste changes. Oxygen has three sources:

- within the product itself,
- ingress through the bottle sidewall, and
- ingress through the closure.

Product may pick up oxygen during the filling process and fillers go to great lengths to avoid this. With respect to ingress through the closure, here PET bottles outperform glass bottles because the injection-molded finish is smoother than the glass molding process can achieve. Nevertheless, about 20% of total oxygen ingress can be attributed to the closure. Many juice products are also sensitive to UV light that can affect color and attract consumers.

Cosmetics

Oxygen ingress can also change the color and olfactory properties of the product. Most companies choose to mold bottles with thick walls instead of other ways of barrier improvement. This has the further advantage that consumers find a PET bottle with near-glasslike properties easier to accept.

Milk

Milk is similar to juice with the added difficulty that milk is not only susceptible to UV but also to visible light. This is not much of an issue with fresh milk that typically has a shelf life of less than 2 weeks and this is not long enough to cause much harm. However, extended shelf life (ESL) milk has an expiry date of 6 months and more and HDPE bottles produced for this market feature a layer of black in a multilayer structure to block all light affecting the product. PET preform manufacturers have taken on that challenge and there is now a process where a black layer is overmolded with PET.

Beer

Beer is in many ways a worst-case scenario for PET. It usually comes in small packages and needs to be protected from both carbonation loss and oxygen ingress. The latter is limited to 1 ppm over 6 months, the highest requirement for any product. Carbonation loss is limited to 10% over the same period, also higher than what most soft drink companies demand. Most beer bottles are produced in multilayer but coating also has found entry into this market.

Chemicals

There is a growing market in the home improvement sector that caters to home owners that take care of their own gardens and cars. Many fertilizers, weed and insect killers, and automotive additives contain volatile chemicals that also tend to migrate through the container walls. Most of these containers are still made in high density polyethylene (HDPE) but there are some in PET as well.

Methods of Barrier Enhancements

There are four technologies currently in use:

- Multilayer
- Coating
- Additives in injection molding
- Monolayer with new materials

Blowing of Multilayer Preforms

Multilayer preforms are created during the injection molding process. A high-barrier material such as nylon or EVOH is injected into the PET melt stream via a secondary extruder, resulting in a three- or five-layer structure (Table 9.5).

Two items to note:

- PET is far superior than the polyolefins HDPE and PP
- While EVOH is the best barrier material, its performance is greatly diminished under high-humidity conditions. Nylon

Table 9.5 Most Commonly Used Materials (Lower Numbers Indicate Better Barrier Performance)

O_2 Permeability	
Material	**Coefficient**[a]
EVA	1990–2780
EVOH (100% RH)	2.16–4.33
EVOH (dry)	0.0276–0.187
HDPE	238–1110
LDPE	1030–1910
Nylon	7.87–11.8
Nylon MXD 6	0.59
PET	22
PP	596–1030

[a] *($\times 10^6$) (cm^3 cm/cm^3 day bar).*

MXD6 has replaced it in many applications as it is not affected by humidity.

The type, location, and percentage of barrier material all have an impact on barrier performance and engineers carefully design packages to suit particular requirements. The process can boost shelf life up to 10-fold depending on the type of product packaged, bottle design and wall thickness, and environmental conditions.

There are currently four processes in use:

- Continental PET's (now Graham Packaging) metering process (proprietary)
- Kortec's sequential process
- Hofstetter process
- Amcor's Amguard process (proprietary)

Graham Packaging's process uses metering pistons that precisely premeasure each quantity of resin before injection. A five-layer structure is formed that also allows the use of up to 35% of postconsumer resin guaranteed not to come in contact with the bottle contents or even the virgin PET (a separate extruder is required though). Graham Packaging uses Sur-Shield as its barrier material. This is a proprietary formulation that acts as an active and passive oxygen barrier, also reducing CO_2 permeation (Fig. 9.44).

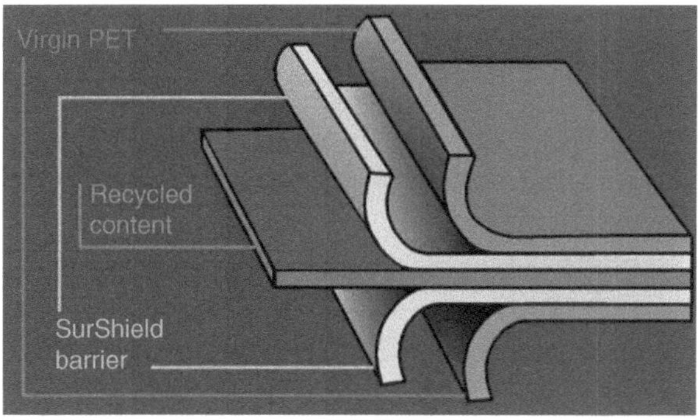

Figure 9.44 SurShield multilayer structure. Diagram courtesy of Owens-Illinois Plastic Group Food and Beverage Products.

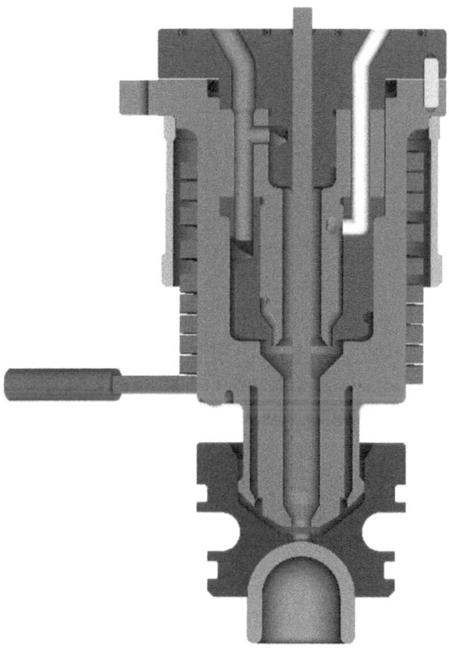

Figure 9.45 Multilayer die. The PET stream in red encapsulates the barrier stream in yellow. Diagram courtesy of Kortec Inc.

Kortec's design is based on a set of "dies" not unlike the ones used in the production of multilayer, extrusion blow molded containers. As can be seen in Fig. 9.45 the PET melt stream is divided into two separate streams. The barrier material—thermally insulated from the PET—enters at the center of the combined PET stream and can be located anywhere in the preform by sequential injection. Often there is no need to have a barrier in the bottom or the neck finish of the bottle and closed-loop control on all injection parameters makes precise barrier placement possible. This method also guarantees that the barrier is in the center of the preform wall, ensuring maximum efficiency of so-called scavenger materials, which are designed to neutralize oxygen.

Proper centering of the barrier also reduces the risk of delamination. When delamination occurs the barrier separates from the PET layer(s) reducing its effectiveness and possibly resulting in an unsightly appearance. To reduce the risk of delamination stretch ratios in both directions must be optimized and preform temperatures should be kept as low as possible (see Chapter 6 on optimum preform temperature). Delamination may also occur when the bottle neck is cut in the so-called blow-trim process and that has prevented this technology to be used in the production of a can replacement (Chapter 9, Section 9.9).

Internal Plasma Coatings

Three companies have made developments in this area:

1. Sidel has a system on the market under the name "'Actis" using acetylene as the gas.
2. Mitsubishi markets a diamond–carbon coating that is similar to Actis but imparts a structure that improves the coating.
3. KHS Corpoplast has developed a competitive system called Plasmax using a silicone oxide as gas (Fig. 9.46).

These coatings are achieved by polymerization of food-safe gases inside the bottle. The gas converts to a plasma state when excited by various forms of energy such as microwaves or radio frequency waves. The plasma then deposits in a layer with a thickness of only 0.00015 mm (0.000006 in.) on the inside of the bottle wall. Thin as this is, remarkable barrier enhancements can be achieved: oxygen ingress reduced to 10–30 times, carbonation loss 4–7 times, and AA migration (Chapter 3, Section 3.6) 3 times compared to a noncoated bottle.

The last point in particular is a definite advantage of internal coatings. AA is a natural substance found in citrus fruits and is also a by-product of injection molding PET. It can add a sweet flavor to the contents of a bottle and both water and beer are sensitive to this. External coatings have no effect on this migration, making internal ones prime candidates for beer. Other advantages are that the barrier layer is protected from external damage and the possibility of using post consumer regrind that will not come

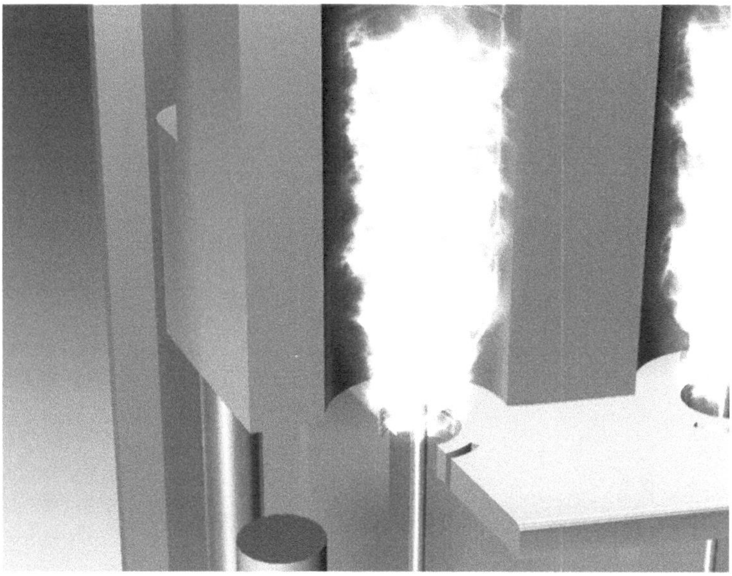

Figure 9.46 Gas changed into plasma coats inside a bottle. Diagram courtesy of KHS Corporplast GmbH.

into contact with the product. Internally coated bottles can also be recycled in the normal recycling stream. One issue that potential users should investigate is the possibility of cracking under specific stresses that might be different in each plant. Actis adds a light yellow color to the bottle that may be unattractive to some consumers but is not an issue with beer that is mostly sold in brown bottles (Fig. 9.47).

These systems use a carousel-type machine that is installed inline with the blow molding machine. Speeds between 10,000 and 40,000 b/h for bottles in the 0.2–0.7 L range are achievable.

External Coatings

Also available are spray coatings and chemical vapor deposition methods. Some are under vacuum, and others are at atmospheric pressure. External coatings may be easier to apply because the outer bottle surface offers better access. However, they must be investigated for possible damage on conveyors or other equipment within a specific production environment, that is, not every plant uses the same type of equipment that might cause damage to the coating. They offer similar performance improvements to internal coatings but cannot affect AA migration.

Figure 9.47 Internal plasma coating system use rotary drives. Picture courtesy of Sidel Inc.

Here are the different solutions:

1. An epoxy-amine coating, marketed under the trade name Bairocade by PPG, has several applications running for fruit juices. This coating must be dried in a special oven system and the bottles stored for 8 h before filling.
2. Several companies are said to be making further developments, addressing the possible cracking of the coating under impact.

Additives in Injection Molding

Both multilayer and coating solutions carry a hefty price tag and many companies are looking for solutions without or with little capital investment even if operating costs may be a little higher. The industry has responded with a number of additives all intended to boost barrier performance.

There are several groups of additives, some of which are only added to the material in preform production while others may also be used in a multilayer structure:

- Oxygen scavengers
- Nanoclays

Table 9.6 An Overview of Available Scavengers

Product	Region	Producer	Resin Base (Format)	Website
Amosorb DFC	All	ColorMatrix	PET (masterbatch)	www.colormatrix.com
Polyshield	Europe	Invista	PET (resin)	www.invista.com
Aegis OX	All	Honeywell	Polyamide blend (resin)	www.honeywell.com
Oxbar	All	Constar	Polyamide MXD6 (bottle)	www.constar.net
Bindox	Europe	Amcor	Polyamide MXD6 (bottle)	www.amcor.com

- Barrier materials with nanocomposites
- Enhanced resins with nylon added at the extruder

Oxygen scavengers are what the name implies: they scavenge oxygen from the product and from the outside. Most of these accomplish this task by cobalt salts that are eager to connect with oxygen. This is called an active barrier against oxygen ingress and does not protect from carbonation loss, which requires a passive barrier. These scavengers can be incorporated into a monolayer material, as part of a multilayer structure, and also into liners of closures. Large converters have their own brands of material mixtures that all incorporate these scavengers (Table 9.6).

BP markets a proprietary PET copolyester under the trade name Amosorb licensing it to various providers. It chemically bonds with oxygen that permeates through the package wall. It can be blended with virgin PET or a barrier material like nylon in a multilayer preform.

Constar and Chevron have developed a nylon enhanced with cobalt salts that acts as an oxygen scavenger and also has superior carbonation retention properties. Marketed under the trade name Oxbar it can be part of a multilayer structure or directly blended with virgin PET. The latter approach reduces the capital costs associated with coinjection systems. Oxbar can also be used in hot-fill applications reducing oxidation of vitamins and so adding shelf life to containers.

The organic components of the scavenger tend to give bottles a slightly yellow tinge that consumers may find unattractive and eventually cause problems in the recycling stream where they could increase the yellowness of recycled material when large quantities of treated bottles enter the stream. A new additive replacing the organic materials with a metal

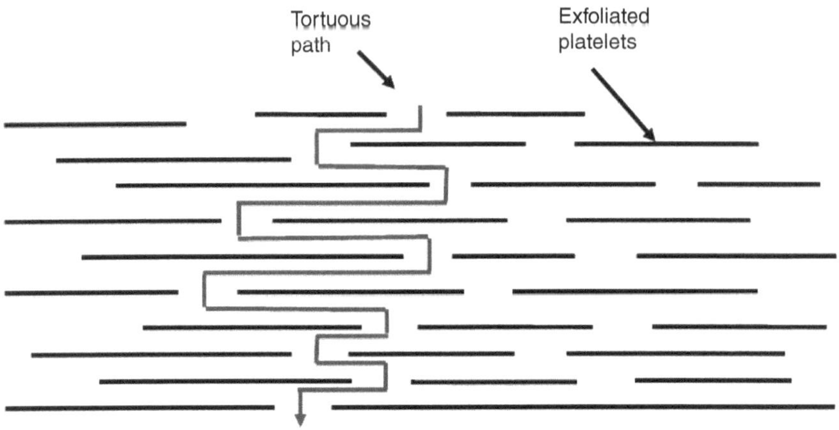

Figure 9.48 Nanoclays present obstacles to oxygen ingress.

catalyst under the trade name HyGuard is currently in production trials. The catalyst converts oxygen into water without discoloration and may become the preferred choice of brand owners. As with every other method the price point will eventually decide the outcome!

Nanoclays work in a different way: they present oxygen from the outside of the bottle with a kind of obstacle course, called tortuous path. The clay platelets are produced to form a high aspect ratio. They are exfoliated into very thin sheets that intersect with PET molecules, impregnable for oxygen molecules. The latter have to migrate around them to be able to penetrate (Fig. 9.48).

Nanoclays are offered under various trade names (Table 9.7).

Nylon can also be added directly to PET with a suitable dosing unit at the extruder throat. Properly disbursed in concentrations of 5–8% it has a similar effect compared to be being enclosed in a multilayer structure. However, the yellowing is more pronounced and it has therefore found many applications with colored bottles such as for beer where this is not an issue. Trade names are Polyshield and Amosorb Solo.

Monolayer Solutions

Both end users and processors are expecting resin producers to develop materials that fulfill all of the enhanced property demands on PET. Polyethylene naphthalate (PEN) is such a material. Manufactured by BP and available under various trade names the enhanced performance of this resin is due to the double-ring structure of the naphthalate molecule. With a glass transition temperature (TG) of 125°C (257°F) PEN provides much

Table 9.7 Selected Nanoclays Offered Under Various Trade Names

Product	Region	Producer	Resin Base	Website
Dnrethan LDPU	Europe	Lanxess	PA6	www.lanxess.com
NycoNano	USA	Nycoa	PA6	www.nycoa.net
Aegis NC	USA	Honeywell	PA6	www.honeywell.com
Nanoblend	Europe	PolyOne	PA6	www.polyone.com
Nanomide	Asia	XanoPolymer	PA6	www.nanopolymer.com
Ecobesta	Asia	Ube Industries	PA6 copolymer	www.UBE.com
Systemer	Asia	Showa Denko	PA6	www.showadenko.com
Imperm	All	Nanocor	MXD6	www.nanocor.com

better hot-fill capability and its low oxygen ingress and carbonation loss are sufficient for most applications with the exception of small beer bottles in hot climates. PEN can be processed on standard equipment albeit running a slightly different process. It is certified for all food applications and can be mixed with PET to form blends of different concentrations tailored to a particular application. However, this has raised issues with recyclability that are still pending.

PEN seems to have all the ingredients of a successful material in the PET world except one: at US$5.50 per kg (US$2.50 per pound) PEN is three times as expensive as PET. Material costs represent a very significant percentage of overall packaging cost, varying with container weight, and PEN can easily double these costs. Therefore, end users have been reluctant to embrace PEN in single-serve applications. However, several multi-serve bottles including beer have been successfully launched in Europe as well as a smaller numbers of single-serve bottles.

The latest addition of the line up of enhanced materials is marketed as "ActiTUF" and produced by M&G Group. It also contains an oxygen-scavenging agent that is only activated when the bottle is filled so that preforms can be stored without sacrificing scavenger performance. Discoloration is an area that is also of concern with this material. ActiTUF has been introduced worldwide and offers excellent barrier enhancements for its users.

Evaluating Barrier Solutions

It is imperative that brand owners test any barrier material they are considering to use. Barrier solution providers may give a barrier improvement factor (BIF) based on the layer thickness or barrier percentage. This is a somewhat theoretical value. It is derived by comparing a standard bottle with one that has the barrier material in a test unit that measures oxygen ingress. Both bottles are first purged with nitrogen then tested daily for oxygen content. This test lasts usually for a few days after which a stable rate of ingress has developed that can then be extrapolated into the targeted shelf life. It does not take into account the real-world stresses that bottles undergo during filling and distribution. As bottles are thus stressed and also expand and shrink under various temperature conditions, microholes may develop that can diminish barrier performance. In other words, a BIF of 10 does not necessarily mean that shelf life increases from 2 to 20 months. In fact it may only increase to 4 months.

To properly test barrier-enhanced bottles a test protocol has to be established that is specific to the product. Bottles should be subjected to the regular filling and handling procedure and then placed in a temperature and humidity controlled environment. These two values should be chosen carefully to reflect what the bottles are subjected to in an average situation. The most effective way of measuring ingress is to place a sensor inside the bottles that measures absorbed oxygen on a continuous basis. At regular intervals color, test, and vitamin content needs to checked and then correlated with the oxygen content in the bottles at that time. Many flavor changes are very subtle and many companies employ professional test tasters for this task, especially in the beer industry. A graphic display given in Fig 9.49 will help engineers understand the performance with different materials and, in this case, at three different temperatures.

There are several conclusions to draw from this chart, which was created with a sensitive juice as product and a standard bottle and one with an oxygen scavenger additive:

- The highest oxygen concentration is in the standard bottle at the lowest temperature, the third largest with the additive bottle at the same temperature. It is apparent that oxygen is picked up by the product itself at the higher temperatures.
- Despite the highest oxygen concentration in the standard bottle at the lowest temperature color (and taste) were not affected. This leads to the conclusion that the interaction of oxygen and product brought these changes and are only slowed down at lower temperatures.

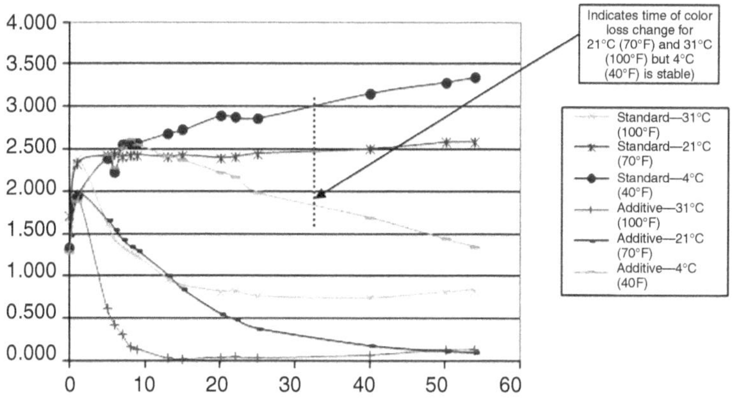

Figure 9.49 Average oxygen concentration post N_2 purge after no. of days.

- The declining curves especially of the two additive-enhanced bottles at the higher temperatures show how the oxygen scavengers work. Tested over longer periods, these curves would have to go up eventually as the scavengers become fully saturated and can no longer attract oxygen.
- The main conclusion the brand owner drew was that this product had to stay in the refrigerated distribution chain because the elevated temperatures are what is causing the product to change color and taste.

Extensive testing like this one is paramount to making the right decisions about any package with sensitive product. Additionally, when considering barrier solutions overall wall thickness should be considered at the outset of any trials. The challenge is to find the economical sweet spot between lightweight and percentage of barrier or coating needed. As weights decrease more (expensive) barrier is needed to make up for the increased transmission rates of the thinned PET wall. Engineers need to go through various combinations of these two parameters to find the solution that best fits with barrier performance and cost.

9.9 Blow-and-Trim Process

Some containers, such as those for coffee, peanut butter, and many kinds of condiments use large neck openings for easy pouring of the product. We typically consider a bottle as wide mouth when the neck size is more than 43 mm. Neck sizes go all the way up to 120 mm. Both the single- and two-stage processes are used, each with its own set of challenges.

The first of these is that wider necks require more room in a given injection platen space. Or, to put it in more economical terms, less cavities can be fitted into the same space, which is costly. As outlined in Chapter 11 this significantly increases the injection machine size requirements if these necks are built into injection tools. In single stage there is usually enough room because the pitch in the injection tool is the same as the one in the blow tool but clamp tonnage has to be increased because the larger neck area requires more of it. Many of these applications are in the small-to-medium sized market and often brand owners cannot find suitable preforms for the two-stage process. The single-stage process has been the historically preferred choice for these two reasons. Machine manufacturers have built special machines with higher clamp tonnage to suit this market segment.

In two stage, there are several challenges starting with feeding the preforms into the machine. All wide-mouth preforms are shaped conically which leads to nesting, causing operation interruptions. As the necks become wider they are increasingly difficult to protect from infrared heat. On top of that blow pressure in the hoop direction increases with the square of the diameter. Both factors combined make it more likely that necks are distorted in the blow process.

There is now a way of getting around these restrictions in the two-stage process and saving tooling costs at the same time.

Instead of building a tool with preforms of a 63-mm neck for example, the tool is built with a longer preform and a 28-mm neck without any thread geometry but with a neck support ring (NSR) or transfer ring. This tool is significantly cheaper because it is the complex neck inserts on standard tools that are the single most costly item. Alternatively, neck inserts from other tools could be used.

Next, preforms are blown in blow molds that include the required neck thread configuration, similarly to extrusion blow molding. A blow dome above the neck merges the 63-mm neck to the 28-mm preform top and bottles leave the blow machine with this dome attached. A further advantage is that blow machines can use standard mandrels and heat shields and neck overheating is not a concern as can be the case with wide-mouth necks (Figs. 9.50 and 9.51).

Bottles are then fed into special machines that trim off these domes and also curl or burnish the resulting edges to various, smooth configurations. These blown threads are significantly thinner than the injection molded ones, thus contributing to an overall weight reduction of the containers that would not be achievable with the standard process. Of course they are also not as strong and may not be suitable for liquids, but there are many applications in powder and piece products where this is not an issue. While blow molds for this process are more complex because of the added

Figure 9.50 Stages from small-neck preform to finished wide-mouth container. Photo courtesy of Belvac Production Machinery.

Figure 9.51 Machine module to go in-line after blow molder finishing the wide-mouth neck. Photo courtesy of Belvac Production Machinery.

neck geometry, only one blow mold per four injection cavities is generally needed and there are still significant cost savings overall. These savings and the ones from the cheaper injection molding tool may be used to pay for the trim machine so that capital outlay overall is about the same for both solutions.

The domes can make as much as 35% of the overall container weight and the economic feasibility of the blow-and-trim process depends a lot on how they are processed after being cut off. Once removed from the container they may be ground and sold as clean flakes. This used to be a losing proposition, that is, buying 35% of the total output for $0.70 per lbs, then selling it for $0.15 per lbs. However, as recycling markets have developed clean flakes can fetch as much as $0.45 per lbs with today's resin prices at $0.85 per lbs. This would increase the resin price to $.99 per lbs, which could still make this proposition a losing one. In order to get the economics on a winning track ground up domes should be used in-house where they substitute material that was bought at full price.

The ideal situation would be to use the ground flakes mixed with virgin PET in concentrations of 1015% on several other machines. This will slightly reduce the IV of the material but have very little if any effect on the properties of the thus produced preforms. If that is not possible and all flakes have to go back into one machine at higher concentrations there can be a variety of issues processors will have to deal with. IV will drop by 0.01–0.03 more than if just virgin PET was used. This may or may not affect preform properties and could be compensated for with virgin PET of higher IV. Flakes will contain amorphous PET from the unstretched neck portions of the preforms and highly oriented PET from the blown portions. The amorphous portions may stick in the dryer while the oriented ones that have crystallization levels of around 25% tend to curl and can mechanically interlock. This may lead to material blockage in the dryer and the hoses connecting dryer to extruder throat. Specially designed dryers and vibrating devices may be needed to avoid problems.

To summarize, here are the parameters to make an informed decision whether this process will lead to more economical container production:

Capital Costs

- Existing or new injection tool; new tool would be cheaper with blow-and-trim.
- Cavitation in existing machine or new injection machine; new machine would be cheaper with blow-and-trim.
- Vibrating devices or special dryers may be needed.
- Blow molds are more expensive with blow-and-trim.
- Trimmer is extra cost.

Process

- Percentage of overall weight of domes, can range from 20 to 35%.

- Usage of ground flakes, in-house in several or one machine or sale of flakes.
- Grinder operation; labor cost mostly as a most companies use a grinder for scrap.
- Trimmer operation; labor plus amortization.
- Heating of preforms in blow machine easier with blow-and-trim.

9.10 CSD Bottle Base Failures

There is hardly a defect that annoys fillers more than the bursting of a bottle during filling because of the mess it makes and the production interruption. For brand owners (if they are different from the filler) it is a leaky bottle in the warehouse or on a shelf that is detrimental to the brand image all companies try to uphold. This failure may relate to the bases of these bottles not being formed correctly or it has more to do with stress cracking agents such as alkali cleaning agents as well as storage conditions (Chapter 11, Section 11.10). There is common ground between these two failures even though they may occur independently (Fig. 9.52).

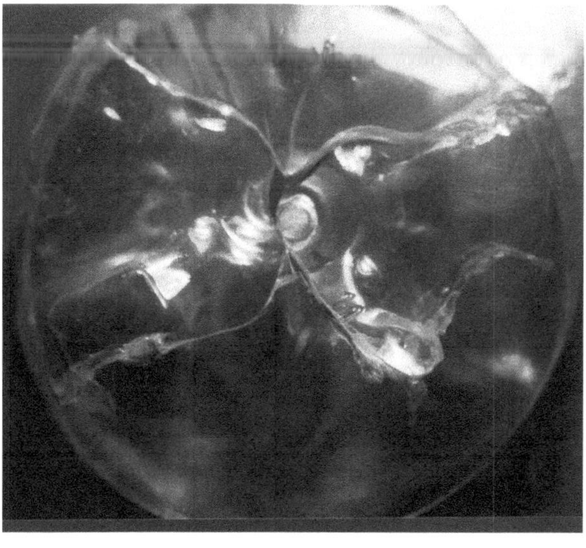

Figure 9.52 The base of a CSD bottle is especially vulnerable to failure.

Preform Production

PET used for CSD bottle production has usually a high IV of 0.80–0.84. The longer chains in high-IV PET hold better together when pressure inside the bottle reaches up to 5 bar (70 psi), a considerable stress only few materials can withstand. IV is tightly controlled during the PET production process but every now and then material may not have the IV it is sold for. This is difficult for the processor to detect, as most facilities do not have the capability to measure resin IV. If low IV is suspected as a cause of failure samples will have to be sent to a lab that specializes in this procedure. It can take days for those results to come back thus further complicating a situation where bottle failures may occur days or weeks after preform production. For this reason it is paramount for the preform producer to keep to good manufacturing practices such as checking the moisture content of dried PET, keeping regrind percentage below 10%, and frequently inspecting preforms for crystallinity and bubbles. Gate crystallinity especially makes the preform base more brittle that can lead to sudden and catastrophic failure.

The most common issue in preform molding is a malfunctioning dryer. The maximum recommended moisture content for PET is 50 ppm, which will lead to an IV drop of about 0.04. (The optimal moisture content is about 30 ppm.) When it goes to 200 ppm the resulting IV drop will be 0.14 resulting in a material with completely different properties. Therefore, dryer maintenance should be a top priority for all PET processors.

A lesser know preform production problem is overpacking the preform by choosing too high of an injection and hold pressure. These pressures should be just high enough to properly fill the preform without sink marks. Increasing the pressure beyond this point leads to additional stress that makes blow molding more difficult and renders the so-produced bottles vulnerable to cracking. It should also be noted that running preforms in free-drop, that is, without a take-out robot and postmold cooling is also not optimal. The long cooling time that is often required on these machines (10+ s) allows shrinkage of the preform in the mold also leading to more stress development.

Pin holes in the vestiges of preforms are a result of poor operating conditions or malfunctioning equipment and may also lead to leaky bottles. Therefore, random preform inspection is a must for all blow molding companies.

Bottle Production

There is still a misconception in some parts of the industry that a heavy base will perform better. Quite the opposite is true. A base for a 20-oz CSD bottle, for example, should be 6–6.5 g and is much more likely to fail when it comes in at 8 g. The reason is that PET becomes stronger with the degree of orientation it receives. Lighter bases are stretched more and withstand stress better. There is always a thicker center in every PET CSD bottle but the key here is to stretch from the corners of this center gradually to the feet of the base so that there is no sharp transition but a high degree of orientation. This is besides the obvious defect when material accumulates in the base because of improper setting of preblow and blow air known as base folding. Insufficient cooling of the base may result in the center disc protruding outward after blowing (the so-called rocker bottom) and is also to be avoided.

There is considerable dispute over how long a preform can be stored before blowing. Is it a week, a month, or 6 months? This depends a lot on storage conditions in terms of temperature and especially humidity. Preforms do age and will absorb moisture when it is present. Free-volume relaxation will also occur whereby the molecular chains move close together as time goes by making the material more brittle. While it is hard to put a fixed time frame on preform storage, keeping them in tightly sealed polyethylene bags and using them within a month is a good practice.

Bottle design also plays a role in failure rates. Over the years many designs have concentrated on reducing the feet radii for stiffness and allowing the center disc to protrude slightly outward. Both measures have helped in making bottles more resilient to failures.

Testing for Stress Cracking

The main test is to subject bottle bases to a high-pH solution and record their behavior. For this purpose a 0.2% sodium hydroxide (NaOH) solution is prepared, the alkalinity adjusted and the temperature set to 22°C (72°F). Aged bottles are then filled with water at the same temperature and pressurized to 5.3 bar (77 psi). The fill level is marked on each bottle after 5 min and they are placed in a stainless tub containing the solution.

Bottles may leak or fail and both occurrences are noted. Brand owners have slightly different bench marks on how many bottles may fail or leak within usually 30 min.

CSD bottles are also subjected to a burst test. In this test bottles are pressurized to 9–10 bar (130–145 psi) for 30 s, then further pressurized to burst. The burst pressure is then recorded and statistical analysis performed on the thus obtained values. No failure may occur within the first pressure stage.

The first test determines the ability of the PET bottle to withstand environmental stress cracking agents while the second verifies that the bottle will not burst during filling and transport. Correct molding of preforms and bottles is the best guarantee against both failures. Very often bottle failures are a combination of factors as noted earlier. A slightly overpacked preform blown into too heavy a base may result in failure whereas one of the two conditions may not have been enough to cause it. It is therefore in the best interest of producers to adhere to all the given guidelines so that one production deficiency can still be tolerated and acceptable bottles leave the factory.

9.11 Recycling of PET Bottles

Plastics contribute only a small portion of consumer waste to landfills (<20%) by weight but double that by volume as hollow containers take up a lot of room and it is room that counts on the landfill site. As cities have grown short of landfill space and the NIMBY movement (Not In My Back Yard) is growing just as quickly as the mountains of garbage people in the industrial world are generating recycling has become an important political issue on the municipal and regional level. The science behind it is sound: According to "Life cycle inventory of 100% postconsumer HDPE and PET recycled resin from postconsumer containers and packaging" prepared by Franklin Associates in Apr. 2007 (available at http://www.container-recycling.org/assets/pdfs/plastic/LCA-RecycledPlastics2010.pdf) "total energy requirements for recycled PET flake are 15–16 percent of virgin PET resin burdens when the cut-off recycling method is used, and 58 percent of virgin resin energy using the open-loop recycling allocation method." The two methods are different in that the former starts with the postconsumer material while the latter attributes a percentage of the "virgin resin burden" to the recycled material (Fig. 9.53).

Recycling rates and procedures vary widely across the globe. The following discussion centers on North America, by far the largest market for PET.

About half of recycling is done by curb side pickup where consumers place recyclable items into blue boxes that are collected with the regular garbage pickup. The goods are then transported to transfer stations where they are sorted into the different streams. For many years, the amount of money municipalities could gain for these goods was in the $40–150 per ton range and the entire operation was losing money. As cities are notoriously cash strapped this situation has caused considerable problems and debates at the municipal level. PET prices had up over the some years and so had the prices for baled bottles and recycled resin so that many of these systems stand on a better footing today. With the recent price drop in virgin PET these programs face again economic problems. Top price for baled bottles is about $550 per ton but can change quickly depending on

Figure 9.53 From flakes to pellets to preforms and bottles. Picture courtesy of Amut S.p.A.

the demand situation. In fact, the uncertainty in pricing and availability is a major factor in hampering the development of new recycling plants.

Table 9.8 shows recycling rates in the North America have been hovering between 20 and 31% with great regional differences but show a slow but steady upward trend over the last years.

All recycling organizations agree that source separation that is part of any deposit system leads to cleaner PET flakes that are then more economical to process, a strong argument for some form of deposit regulation.

According to the latest data available (2014) the largest percentage of recycled PET (41%) is going into the fiber market to end up in carpets and clothes. However, Table 9.9 shows a worrying drop of recycled polyethylene terephthalate (RPET) use of bottle-to-bottle applications from 2013 to 2014 that now ranks only third behind fiber and sheet/film with a 22% share of the overall RPET market. This is disappointing as recycling back to bottles makes the most out of the initially invested energy that went into virgin resin production and also sells for the highest price.

Bottle-to-Bottle Recycling

There are now a number of ways to convert postconsumer waste into food grade bottles or sheet. The starting point is always baled bottles,

Table 9.8 Recycling Rates in the United States

Year	Total US Bottles Collected (mm lbs)	Bottles on US Shelves (mm lbs)	Gross Recycling Rate (%)
2004	1003	4637	21.6
2005	1170	5075	23.1
2006	1272	5424	23.5
2007	1396	5683	24.6
2008	1451	5366	27.0
2009	1444	5149	28.0
2010	1557	5350	29.1
2011	1604	5478	29.3
2012	1718	5586	30.8
2013	1798	5764	31.2
2014	1812	5849	31.0

Courtesy: NAPCOR.

already presorted into PET only and sometimes in clear and green colors. The first step is clean flakes that may be directly processed in specially equipped extruders or solid stated in specially designed reactors.

Bottles always come in tightly packed bales that are held together with steel wire. They are grabbed from the sides with specially equipped forklifts. Baled bottles contain a variety of surface contaminants such as labels, ink, closures, other plastics such as PVC, residual content, aluminum and steel cans, and plain garbage that was somehow mixed up in the recycling stream. Bottles must be singulated out of the bales so they can be sorted and this is the task of the debaler. Usually a large, rotating drum chews layers of bottles from the bales dropping them onto conveyors. In some installations these bottles are first washed to get the outside clean which assists in sorting. All plants use automatic sorting systems and some form of manual operation that may happen before or after the automatic sorting. PVC bottles and labels as well as aluminum closures on plastic bottles are the banes of all recycling plants. X-ray scanners do a reasonable job at sorting PVC bottles out but have a harder time with thin PVC labels. Aluminum closures may be caught in a separate metal-sorting operation that is often done on washed flakes but all systems only work to a certain percentage of accuracy with the result that washed flakes will contain small quantities (<10–40 ppm) of PVC and aluminum.

After sorting bottles are grounded into flakes. Dry grinding has the advantage that thin label parts or barrier resin from multilayer bottles are

Table 9.9 RPET Product Categories 2004–2014

Product Category	2004	2005	2006	2007	2008	2009	2010	2011	2012	2013	2014
Fiber	479	463	422	383	391	344	381	398	512	558	638
Sheet and film	58	71	74	128	153	159	195	202	307	315	365
Strapping	116	131	132	144	137	114	127	120	136	140	126
Engineered resin	12	8	9	11	7	10	9	See other	See other	See other	See other
Food and beverage bottles	126	115	139	136	141	203	216	242	276	425	351
Nonfood bottles	63	63	49	60	55	65	58	57	50	50	57
Other	24	13	30	38	31	42	16	21	31	25	27
Total converter consumption	878	864	855	900	915	937	1002	1040	1312	1513	1564

Courtesy: NAPCOR.

easier separated via elutriation systems. These work by blowing a low-pressure air stream upward as ground flakes fall down. The lighter parts then get caught in the air stream and can be discarded. Wet grinding on the other hand has the advantage that flakes move dynamically through a water stream which improves washing efficiency. Blades also last longer (up to 1 week) as they are constantly cooled during grinding.

Caustic soda or citric acid may be used to wash residue, especially glue off whether the bottles are ground dry or wet. Recycled flakes are also slightly gray and many recyclers do not sort out the light blue colors of many bottles as a certain amount of blue clears up the slight haze in the regrind. After washing flakes are dried and often go through a metal separator that discards both ferrous and nonferrous particles.

Polyolefins (mainly HDPE and PP from closures) have a lower density of 0.9–0.96 g/cm^3 than PET (and unfortunately also than PVC) at 1.33 g/cm^3 and are removed in sink-float tanks where the light parts float to the top and are removed. They are then sold to companies that use them in parts where color variation is not an issue such as black pipes. The washed PET flakes may undergo further processing (as shown later) or are sold to companies that convert them into fiber, strapping, film, or sheet. Green bottles are usually sorted out at the beginning of the process and are sold as green flakes for various applications.

As PET is processed into preforms about 0.02–0.04 of the IV is lost in a well-run process, more if the material is not processed correctly. Bales of postconsumer bottles may contain CSD bottles made from high IV (~0.82) resin or water bottles made with resin of much lower IV (~0.72). Bottle IV then ends up at around 0.78 and 0.68, respectively. The user of the flakes cannot be sure what the IV of the flakes is as it depends on the percentage of water and CSD bottles in the bale and how well the preforms were processed. Users may employ a variety of secondary processes some of which allow the setting of an IV target. These installations also decontaminate the material to food-grade quality. While most of the surface contaminants have been removed during grinding and washing but toxins such as motor oil, solvents, or pesticides may have penetrated the bottle walls and must be removed to very low levels in order to obtain food grade suitability from the governing regulatory agencies. Threshold is <10 ppb (parts per billion) daily intake of all contaminants; depending on migration behavior from packaging to content the level in the wall is limited to a defined amount (ppm). Another issue that can be addressed is the reduction of AA, an important parameter for water bottle production (Chapter 2).

The US Food and Drug Administration has issued nonbinding guidelines that stipulate a level of 220 μg/kg residue of all contaminants for PET bottles with a wall thickness of 0.5 mm (0.020 in.). This value would increase for bottles with less wall thickness. It recommends conducting a "challenge test" whereby a cocktail of known contaminants are deposited in a virgin PET bottle and left for 2 weeks at 40°C (104°F) with periodic agitation. The FDA proposes the following mix of chemicals (one of each group):

Volatile Polar	*Nonvolatile Polar*
Chloroform	Benzophenone
Chlorobenzene	Methyl salicylate
1,1,1-Trichloroethane	
Diethyl ketone	*Nonvolatile nonpolar*
	Tetracosane
Volatile Nonpolar	Lindane
Toluene	Methyl stearate
	Phenylcyclohexane
Heavy metal	1-Phenyldecane
Copper(II) 2-ethylhexanoate	1-Phenyldecane
	2,4,6-Trichloroanisole

The prepared bottles are then rinsed and undergo washing and decontamination within the proposed recycling process. The resultant RPET should be analyzed for residual contamination. This is of course a worst-case scenario as one can assume that only a small percentage of baled bottles are contaminated in this way.[c]

The tasks for equipment downstream from flake washing can be summarized as follows:

- Decontamination
- IV control
- AA reduction

Some of the flakes may be converted to pellets but decontamination can also be done on flakes. In either case heating under vacuum lowers the boiling point forcing contaminants to the surface of flakes or pellets where an inert gas (usually very pure nitrogen) removes them. During the heating AA is also removed. Theoretically, these flakes could be used directly in an injection molding machine to make preforms but are more commonly

[c]Footnote: Food and Drug Administration, "Guidance for Industry: Use of Recycled Plastics in Food Packaging: Chemistry Considerations, http://www.fda.gov/Food/GuidanceComplianceRegulatoryInformation/GuidanceDocuments/FoodIngredientsandPackaging/ucm120762.htm, Aug. 2006

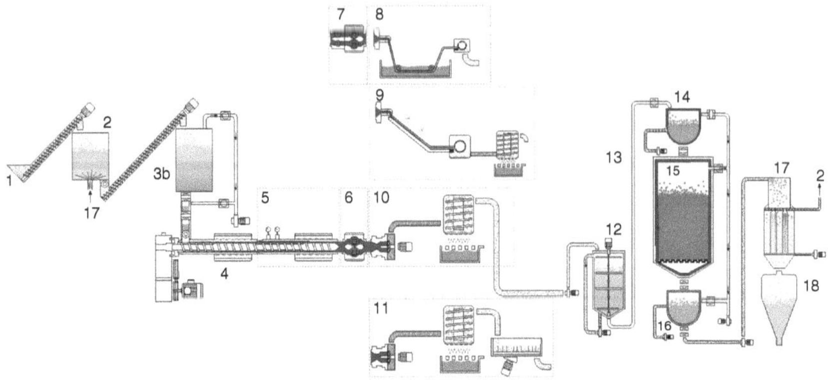

Figure 9.54 Solid state polycondensation. *1*, Screw conveyor; *2*, hot air drying unit; *3b*, predrying unit (vacuum); *4*, extruder; *5*, high-vacuum degassing extruder; *6*, melt filter w/o back flushing; *7*, melt filter with back flushing; *8*, strand pelletizer; *9*, automatic strand pelletizer; *10*, underwater pelletizer; *11*, underwater pelletizer; *12*, crystallizer; *13*, vacuum transport with in-line crystallizer; *14*, preheating unit; *15*, SSP reactor; *16*, cooling unit; *17*, energy recovery kit; *18*,. storage silo. Diagram courtesy of Starlinger.

used to produce sheet for items such as egg cartons or products with a very short shelf life. The processing of flakes in injection machines is a challenge because of the lower bulk density of the flakes and their propensity to mechanically interlock as they are dried and the thin sidewalls tend to curl. This behavior may lead to bridging whereby clumps of dried flakes block the extruder throat. Specially designed driers and hoppers are meant to prevent this problem from happening. These extruders need to use melt filtration to remove left-over particles and coupled with the low bulk density of the flakes this may lead to surging problems. New developments with twin-screw extruders are an attempt to get around some of them and the future may very well be in this process.

However, currently most users opt to convert flakes into pellets in extruders specially manufactured for this purpose. While pellets produced in this way may be used for a variety of applications further solid state polycondensation (SSP) allows the exact setting of a required IV value, further decontaminates the material, and so delivers the best quality possible. Fig. 9.54 shows how various elements may be put together.

SSP can also be performed on flakes to create high-value RPET. By employing dryers/crystallizers under high vacuum in batch mode, washed flakes can be dried, purified, and IV-controlled before they (optionally) enter the extruder that converts them into pellets. Special melt filters remove leftover particles to 32 μm and the resulting melt can either be directly

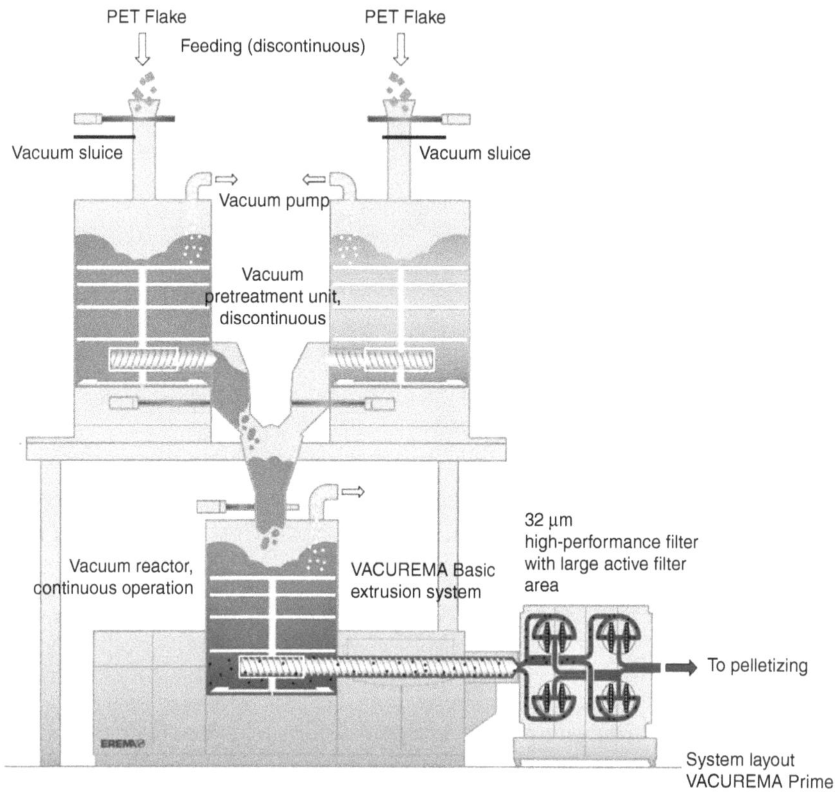

Figure 9.55 This installation controls contaminants and IV by performing solid-state polycondensation on flakes rather than pellets. Diagram courtesy of Erema Engineering Recycling.

used for the production of preforms, sheet, or strapping or fed to a pelletizer. A diagram of this process is shown in Fig. 9.55.

It is apparent that users have a number of choices in how to process washed flakes. Careful consideration of the consumer application is paramount in choosing the optimal solution.

9.12 Preform Aesthetics in the Two-Stage Process

One of the main advantages of the single-stage process is that preforms stay untouched in the machine and therefore bottles have no marks distracting from their brilliant appearance. The same cannot be said for the two-stage process. Here preforms fall onto conveyor belts and from there

Figure 9.56 This mark can be felt on the inside of the bottle and is caused by preforms hitting each other or other parts of the conveying process.

into storage devices. Next they are dumped into the blow machine hopper, go again onto conveyor belts and are unscrambled. As a result preforms receive little scratches. Most of these happen while they are still warm. They are then magnified during blow molding as each part of the preform is stretched multiple times. When holding blown bottles against a light source many scuff marks can be seen. Large labels or intricate patterns on the bottle help in masking these defects.

There are of course a number of aesthetic defects that stem from the manufacture of preforms such as gate crystallinity or burn marks. These have to be corrected in the injection molding process. Typical two-stage defects are touch marks. These can be on the outside or, somewhat surprisingly, on the inside of the preforms. This is because as warm preforms hit each other the impact transfers through the preform and forms a little dimple inside (Fig. 9.56).

There are however a number of measures processors can take to minimize if not eliminate these effects. Lowering the demolding temperature is one of them and rather than increasing cycle time processors should follow these guidelines:

- Cooling water temperature should be kept at 10°C (50°F). This requires dehumidifying the plant or machine during the summer months.
- A pressure drop of 5 bar (70 psi) between the in and out of the water cooling circuit. This ensures turbulent flow rather than laminar flow and leads to better cooling.
- Preforms should be kept in the take-out plate as long as possible. Depending on the overall cycle time the turning of the take-out plate and drop-off of preforms should be delayed as much as possible.

- Injection machine manufacturers have gone to great lengths to overlap machine functions in order to save cycle time. For example, applying clamp pressure can be done at the same time at the start of injection because there is little force on the clamp while the cavity is still empty. Additionally, clamp tonnage can be released during cooling time after the hold pressure has been relieved. Processors should use these functions to their full extent as they are often offered as options.

The configuration of conveyors and storage placement is also important. The exit conveyor should be a soft-drop type conveyor. This means that there are no rollers or other hardware underneath the area where the preforms are dropped minimizing banging. Instead of dropping preforms into gaylords where the first to enter drop over 1 m (3 ft.) in height, there are now a number of devices that aim to minimize the drop distance and move preforms more gently toward the storage container. Bucket and spiral systems seem to be at the forefront of these efforts (Fig. 9.57).

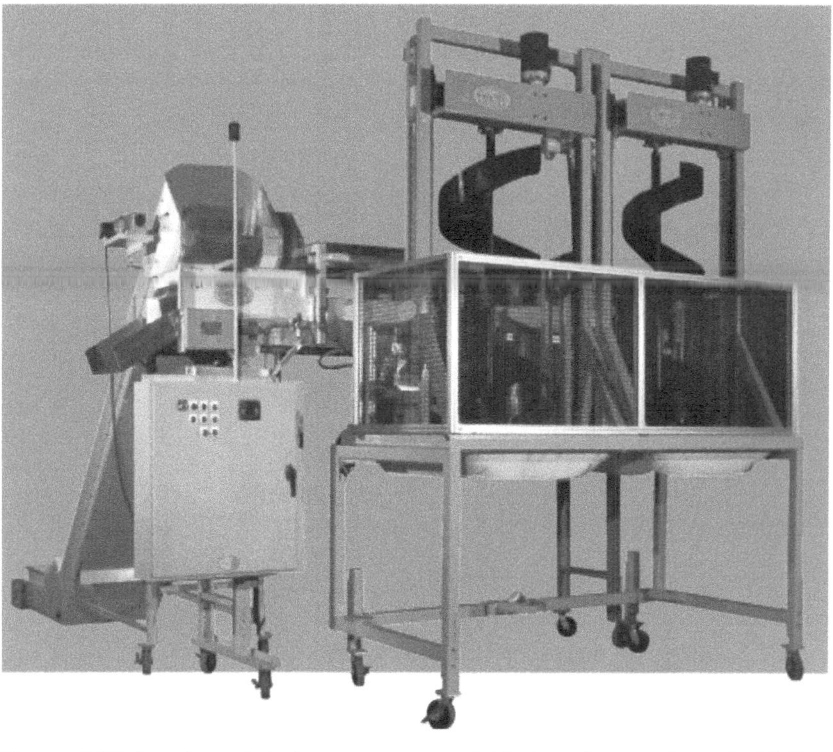

Figure 9.57 A variety of devices are now available to soften the impact of falling preforms and thus reduce the number of scuff marks. Picture courtesy of Han-Tek.

9.13 Blowing Thick-Walled Preforms

Typical preform wall thickness is 2–4.5 mm (0.080–180 in.). This allows typical reheat times of 15–30 s. For containers in sizes of 5–20 L preform wall thickness increases from 6 to 9 mm. These preforms take much longer to reheat as it takes time for the infrared heat to penetrate the thicker walls. The ovens also need good airflow to prevent the outside surface from overheating.

A very common category in this segment of the industry are the 5 gal reusable containers that are placed onto water coolers in many offices and homes. Traditionally, these were made by extrusion blow molding with an integrated handle out of polycarbonate (PC). PC is an ideal material for this application. It can withstand the heat from the washing process as its glass transition temperature is 125°C (257°F). It also has great wear resistance. These containers are often slid across racks on delivery trucks causing significant wear. The major disadvantage of PC is its high cost, about 3 times that of PET.

For this reason bottlers have looked to cheaper alternatives. A few have opted for extrusion blow molded PVC but the majority is using PET. Many 5 gal bottles are made on semiautomatic machines. This is because smaller water bottlers do not need high-output machines and buying the containers from a converter is expensive, not the least because the shipping charges can outweigh the container costs (Fig. 9.58).

Figure 9.58 Preforms with walls thicker than 5 mm need extra time in the oven system.

One advantage of the extrusion blow molding process is the ability to incorporate a handle that makes lifting the full container much easier for the consumer. Therefore, many PET bottles of this size are molded with an inserted, injection-molded handle. It is placed into the mold before blowing and the preform inflates around it. This requires the preform to be quite hot, in the area of 110°C (230°F) so that it will find its way into the small crevices around the handle. As outlined in Chapter 2 this is very close to the crystallization temperature of PET and therefore precise heating is required. This in turn requires that the speed of the oven chain that drives the preforms through the ovens is precisely controlled. Not all semiautomatic machines have this control and buyers should be insisting on a digital control mechanism rather than a simple dial. An advantage of semiautomatic machines is that the oven chain is moving continuously. It is therefore possible to make containers with good consistency.

Another approach to the handle problem is to injection mold it in the preform itself. This requires the single-stage process because it will be very difficult to reorient the somewhat awkward preforms. Fig. 9.59 shows a picture of a successful application of this idea.

Figure 9.59 These 20 and 50 L containers have injection-molded handles that were part of the preform design. Photo courtesy of CyPET Technologies.

Thick-walled preforms are also molded on automatic machines and several companies now offer machines with continuous motion of the preform. No rotary machine exists for this purpose so linear machines are the only choice. The feeding of the rather unwieldy preforms can be a challenge and requires some special mechanism. While a semiautomatic machine will yield about 50 pieces/h fully automatic machines can bring that about 500 pieces/h. The limit here is that it takes extra blow and exhaust time to first inflate and then exhaust the large volume of air required.

Large containers can also be made in the single-stage process. One disadvantage is that, even though the blow process takes longer than usual, the injection process takes even much longer. It is necessary to cool down the thick preform wall below the point of crystallization and as we have seen, the necessary cooling time is largely dependent on the wall thickness. The machines are therefore rather slow. But the long injection time can also be used to some advantage as it is possible to delay blowing to the end of the injection cycle. This gives the preform wall temperature time to even out and make the container wall thickness more even. Also, special preforms are more difficult to obtain and making them yourself can for that reason alone be a winning proposition.

10 Troubleshooting of Blowing Problems

Chapter Outline

Stretch Blow Molding. http://dx.doi.org/10.1016/B978-0-323-46177-1.00010-X

This section gives general guidelines on how to improve one's skills in managing the blow process as well as specifics on common problems.

10.1 General Guidelines

One of the challenges in successful stretch blow molding is to correlate heater lamp settings with preform temperature and bottle wall thickness. Even though the preform may have an even wall thickness, this does not mean that all lamp settings will be the same to achieve an even temperature. And an even preform body temperature is the best base to work from to achieve even bottle wall thickness. If the preform has an even body temperature with a base temperature slightly lower, the preform design must be suited to the bottle in question as changing blow parameters will not have enough effect to force material into areas of the bottle that may need it. In some cases it becomes necessary to reheat the preform to an uneven temperature profile in order to achieve this. Either way processors generally will not know what the temperature profile is since they have only one temperature sensor along the preform wall.

Temperature in any particular part of the preform relative to other parts depends not only on the lamp setting but also on the preform wall thickness and position in the oven system. Air temperature is higher toward the top of the oven and the distance between lamp and preform is not the same for every lamp.

For these reasons it is of paramount importance that processors understand the effect each lamp has on the bottle wall thickness. Here we describe a way to accomplish this: place a cold preform on a mandrel inside an oven. Measure the distance between the underside of the neck support ring (NSR) of the preform to the center of each lamp. Mark lamp locations on the preform by scratching the surface with a steel ruler at the distances measured. Do not use a marker pen to do this as the color changes the reheat characteristics of the polyethylene terephthalate (PET)!

To help accomplish this task a special pipe may be used. The pipe has an inside diameter large enough to accommodate the preform but small enough that the NSR bottoms out on the end of the pipe when the preform is inserted. Cuts made by a band saw about 1/3 of the way through the pipe are at the lamp distances. A steel ruler can now be swiped across the preform inside the pipe marking it. The marks will be barely visible but as they are blown up in the bottle they will easily be recognized (Figs. 10.1 and 10.2).

Next, blacken the necks of these scratched preforms so that they can easily be identified after blowing. Now blow and examine the marked bottles. The scratches will be visible when holding the bottle against a light

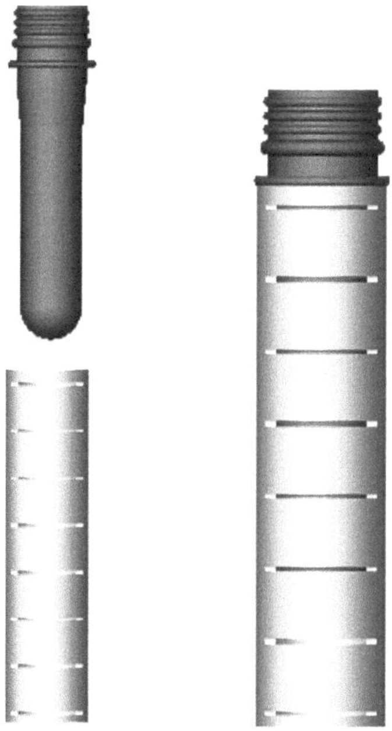

Figure 10.1 A special pipe facilitates the marking of the preform. Once the preform is inserted, it can easily be scratched with a steel ruler.

Figure 10.2 Preform with scratches showing location of lamps.

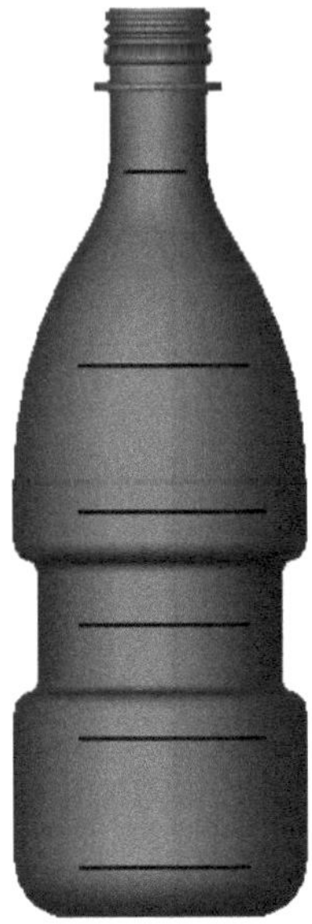

Figure 10.3 Scratches enhanced with marker indicate location of lamps.

source. Mark the original scratches with a permanent marker for easy examination (Fig. 10.3).

After blowing some initial bottles it is suggested that bottles be cut lengthwise and the wall thickness measured. If a device with a Hall effect sensor (known under the trade name "Magna-Mike") is available, cutting can be omitted. Hall effect sensors are very accurate and have greatly simplified this task. Ultrasonic measuring devices on the other hand need calibration for different wall thicknesses and are less reliable though still usable.

Using the wall thickness distribution data that has been gathered, and lamp positions identified, lamp settings can be changed as needed. Thicker areas require more heat and vice versa. However, always keep in mind that the preform blows as a whole. This means that every part of the preform affects every other part and the base has a special function in this behavior. Because the stretch rod enters the base of the preform first and starts pushing on it, temperature, wall thickness, and structural integrity (i.e., the shape of the base) all work together and determine how much the base stretches before other parts of the preform start stretching.

As mentioned earlier in Chapter 8, Section 8.2 PET always stretches at the weakest point. During stretching, the stretched area undergoes molecular orientation, increasing its strength. Now other areas of the preform are weaker relative to the stretched area and the stretching action moves to the part that is now the weakest. All of this happens at very high speeds, but the effect is no less profound. Unlike other plastics, PET does not blow out easily because of the exponential increase in tensile strength in highly stretched areas. PET has been successfully stretched into a wall thickness (wall thinness would be a better word) of only 0.05 mm (0.002 in.), less than ordinary paper. It is a common mistake to want to add material to areas that are not performing in bottle tests. In most cases blowing these areas slightly thinner is all that is needed!

Only when there is a large enough temperature difference in the preform itself, is it possible for certain areas to become very thin while others remain thick. Identifying the responsible lamp and reducing heat in the thin area is the most promising approach. More on this can be found in later sections.

Another way of gaining an insight into the process is to make preblows, also with marked preforms, and examine the inflation. On most machines the high-pressure air can be turned off. It can be helpful to know how much the preblow pressure inflates the preform. Consider also the fact that some of the preform bubble has shrunk back by the time it is removed from the mold, so the inflation process is actually a little further ahead than the half-blown bottle shows (see Chapter 7, Section 7.3 for a more detailed explanation).

Once a process has been established, section weighing is an important tool that is very popular with carbonated soft drinks (CSD) bottles but can be used for any bottle. It is usually done with hot-wire cutters. Wires may be mounted at different heights on a table or in a fixture. High current/low voltage power runs through the wires heating them enough to easily cut through a PET bottle. The bottle is cut into three sections: neck, body, and base. It can also be useful to cut off the unstretched neck portion of the bottle as a fourth section. Once a process that yields well performing bottles has been established, these weights are noted and checked as part of the quality procedure. Deviations in these weights indicate a shift in the process that might require operator intervention (Fig. 10.4).

10.2 Starting a New Process

It is paramount that all preforms are at the same temperature before they enter the blow machine. Preforms should settle 24 h after injection (see exception to this rule in Chapter 9, Section 9.6) and if stored outside they should be brought into the factory to allow for possible temperature differences to even out. The amount of time depends on the difference between

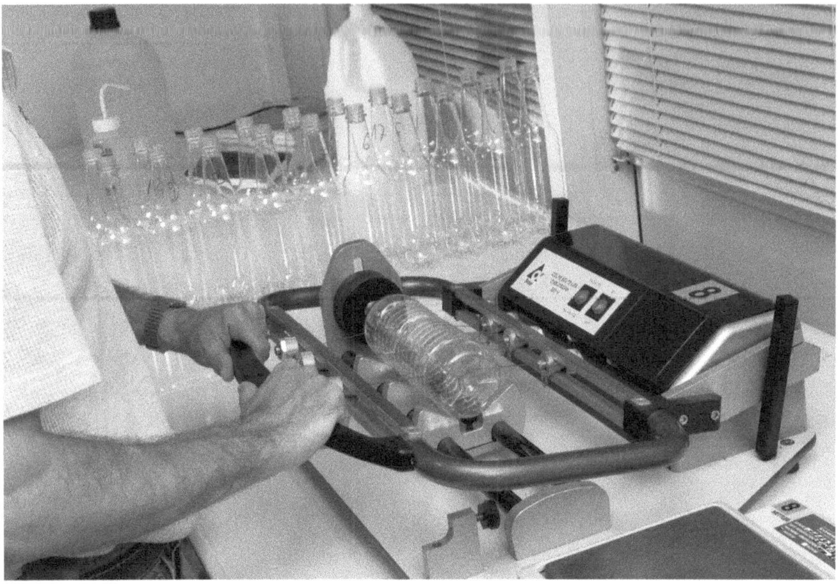

Figure 10.4 Weighing sections of blown bottles separated easily with this device allows monitoring of the process. Photo courtesy of AGR International.

the inside and outside temperature. Loading preforms at different temperatures may lead to haze or pearlescence, depending on which preform the process has been set up.

First the preform must be examined and the lamps adjusted according to experience with similar ones. It is virtually impossible to give a heat profile that will work in any machine and with any preform. Machines may have one or up to 24 ovens and the preform wall thickness may be 2 or up to 9 mm. As explained earlier lamps #1, and possibly #2, will be at a high setting of 90% or higher. It is advisable to run lamps no higher than 95% during start-up in order to leave some upward room for fine tuning. Lamps pointing to the center of the preform need to be set lower because heat from the bottom and top lamps also radiates toward the center.

If throughput rate and wall thickness allow, it is advantageous to have other lamp settings high as well and turn some of them off instead of running all of them at a lower setting (Chapter 7, Section 7.4). In Fig. 10.5, lamps 3, 4, 5, and 7 would be candidates for this. (Doing so creates some new challenges in the process and should only be attempted by experienced processors when a particularly difficult bottle must be blown. The challenges arise from the interrelationships of lamp radiation and only experimentation will tell which lamps can safely be turned off.) Preform cooling should be set to a medium value. The exact amount of fan cooling can only

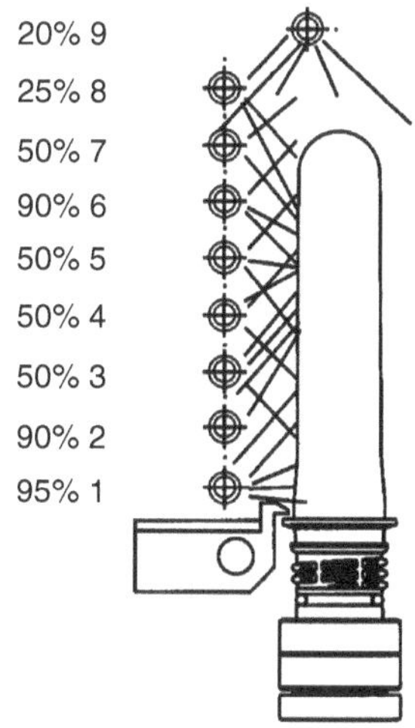

Figure 10.5 Possible start-up heat profile for a preform with even body wall thickness. Diagram courtesy of KHS Corpoplast.

be determined when the machine has been blowing bottles for some time. It can take up to 15 min for the blow machine to settle into a process from a cold start or as little as 30 s, depending on the fan speed and oven temperature. When oven temperature is monitored it is done so by a thermocouple in one oven module. On machines that allow turning individual lamps on and off, this can lead to some perplexing setup problems. This is because if an operator turned some or all of the lamps in this particular oven module off, he or she will have recorded very low oven temperatures that were not typical for the setup. If on the other side, a tall preform is blown with the lamp closest to the thermocouple turned to a high value then the reading will be very high. This tells us two things: the actual oven temperature reading should not be taken at face value; we should always check where the thermocouple is located and judge from that position whether the oven is "hot" or "cold." Second, setup sheets need to keep track of which lamps are turned on and off for each oven.

The process is started by feeding two preforms per cavity and letting them reheat and blow. Lamp settings at this point should be even higher than the anticipated level during production because the system is still cold. Or we can set and wait for a certain oven temperature before blowing with the considerations above in mind. It can be helpful at this point to reduce the air pressure to 20 bar. It is sometimes a real challenge to get the process to the point where every preform blows into a bottle without rupturing at some point. Bottles may rupture at the base or along the body. In the former case the preform is generally too cold or there is a combination of warm base and cold preform body. Examining, at the point of breakage, whether the base stretched before it ruptured can give a pointer in the right direction. When the preform wall ruptures it may be overheated in that area but cannot stretch enough because other areas are too cold. Knowing the relationship between lamp position and bottle wall can be an invaluable help here.

To avoid rupture problems altogether it is possible to run preforms through the machine without applying any blow air. Examining the reheated preforms can reveal their temperature. By squeezing them manually the preform walls should come together with the inside walls touching, but return to their original position out when released. If it is not possible to squeeze the walls together the preform is too cold. If the insides stick together, the preform is too hot. Using the infrared temperature read-out as a guide, it should be noted that this reading may be 10°C–25°C(18°F–45°F) higher or lower than the actual preform temperature. Never leave the machine running with heat but without preforms! The mandrels could heat up to a point where enough heat is conducted into the preform necks to deform them. Modern machines reduce the lamp setting to a set value (usually around 40%) whenever no preforms are present.

It may happen during start-up, or even at other times, that the machine stops with heated preforms in the blow molds. These preforms sometimes stick to the stretch rods making it very difficult to restart the machine. Some machines include a special blow pressure setting that is activated during the clearing of the preforms out of the blow molds. A short shot of high-pressure air is usually sufficient to blow the preforms away from the stretch rods. Activating the blow air valve manually can compensate for the lack of this feature on some machines.

Once a process has been established that produces bottles, the operator should check the bottle wall thickness and make lamp adjustments as discussed earlier. Once the wall thickness is fairly even, it is time to check the oven temperature and decide whether it is possible to lower the temperature. Of course, this is only necessary when blowing bottles with high stretch ratios and/or high quality demands. Whether it can be done will depend on the highest lamp setting and whether all lamps in that row are turned on. Processors should now determine whether they can lower the preform temperature without encountering pearlescence. As explained earlier (Chapter 7, Section 7.5) preforms blown at the lowest possible temperature yield the best performing bottles. Increasing the fan cooling while checking bottles for pearlescence is the best approach here. When reducing oven temperature and bottles show no signs of pearlescence, then wall thickness should be checked again. The row of lamps responsible for the thinnest wall is now being identified and lamps in that row can sequentially be turned down or individual lamps turned off.

Once pearlescence starts to be seen, the operator can check wall thickness again and either reduces oven cooling or turn lamps slightly higher in areas with the thickest wall section. Once pearlescence has disappeared, the process can now be regarded as producing the best quality bottles from that particular preform.

10.3 Preblow Pressure Control

The correct preblow pressure setting depends on preform temperature, distance of preform wall to bottle mold wall, and preform thickness.

The goal is to blow a "bubble." This bubble is too big if it touches the bottle mold wall and too small if it still shows parts that are almost as small as the preform.

Most machines allow a so-called preblow audit. With this feature active, the machine only blow with preblow pressure and so allows the operator to see the effect.

Fig. 10.6 is a good illustration of what the effect of preblow pressure should be, it is also an illustration that this method is prone to significant errors: the

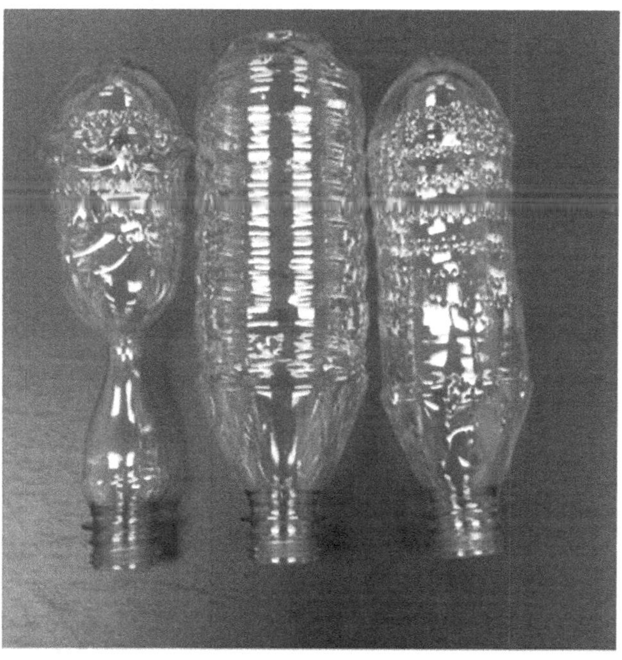

Figure 10.6 Three bottles made with preblow pressure only. Left is too little, middle is too much, right is the right preblow pressure.

three bottles are three consecutive preblow audit samples off a rotary machine. Small changes in preform heat and response time of the preblow valve lead to significant changes in preblow audit results especially when preforms have been heated up to well over 100°C (212°F) in order to use stretch ratios above 12. Another source of inconsistency is the air volume flow control to each cavity. This is typically adjusted manually and may be different.

It should be obvious that there is no fixed, correct value for the preblow pressure setting. A preform at a temperature of 88°C (190°F) needs a lot more pressure to inflate than a preform that is 110°C (230°F) hot. This means you may have to redo the preblow audit when you change the preform temperature. Values can range from 4 bar (60 psi) to 20 bar (290 psi).

Single-stage machines typically do not have a preblow audit feature. You may have to manually turn the feed to the high-pressure valve off to achieve the effect. Adding a manual shut-off valve after the high-pressure tank may be the only way of doing this if it is not already there.

10.4 Changing Preform Temperatures

This is very different in two- and single-stage. In two-stage machines preform temperature is determined by the lamp settings and fan speed as well as the residence time of the preforms in the oven.

When you want to make a change of overall temperature without changing the profile you should keep in mind that 2% of 30% is a different change than 2% of 90%. Let us say your lamp settings are like this:

Lamp #1	88%
Lamp #2	73%
Lamp #3	52%
Lamp #4	45%
Lamp #5	40%
Lamp #6	42%
Lamp #7	35%
Lamp #8	58%

Some machines allow control over an overall oven setting, which makes this task very easy. If this is not the case it requires a little bit of math to accomplish. If you wanted to add heat without changing the profile you would need to make the same percentage increase for each zone. A 5% increase would mean about 2% for lamp #7 but over 4% for lamp #1. (You can calculate this by multiplying numbers with each other: $35 \times 0.05 = 1.75$.) This does not have to be completely precise but in this case I would increase lamps # 4, 5, 6, and 7 by 2%; lamps #3 and 8 by 3%; and lamps # 1 and 2 by 4%. This will keep your profile intact but give you a higher preform temperature.

In single stage we control preform temperature by a number of factors:

- barrel, hot runner, and nozzle temperature
- injection speed
- hold time and to a lesser extent pressure
- cooling time
- conditioning time and temperature (when available)

Decreasing hold and cooling time reduces cycle time but increasing them adds to time and so makes the product more expensive to manufacture. Therefore, if a lower preform temperature is required, lower temperatures should be the first measure if they are above 275°C (530°F).

Injection speed changes the temperature profile of the preform and slowing it down to give the material more time to cool is generally not the best idea. Use hold and cooling time instead.

Conditioning is more suited to cool certain areas rather than the overall preform and is therefore only partially available for our purpose here.

10.5 Output Control

These considerations are very different between two stage and single stage. Two-stage first: each machine has a mechanical maximum speed. For a rotary machine it is how fast the blow wheel can turn. For linear machines it is the time it takes to bring preforms in and out of the clamp area and the opening and closing of the mold. Often machines are listed as being able to run so many bottles/cavity per hour. This is somewhat misleading because it does not reference process time, that is, preblow, blow, and exhaust time. The listed output always refers to lightweight water bottles of 500 mL or 20 oz volume without telling us what process time these outputs leave.

The fastest rotary machines now run 2400 bottles/cavity per h. That means they make one full turn in 1.5 s. About half of that is process time; so don't think you can run a 21-g CSD bottle at that speed!

Besides the mechanical performance, there are other factors that may limit output:

- Cooling time required to cool the bottle base enough so it won't shrink out.
- Oven capacity may not be able to heat very thick preforms in the time it takes to cool the base.

- Very large bottles such as 5 gal containers take considerable blow time to fully inflate.
- Large containers also need longer exhaust time.

As an operator you should first determine the minimum cooling time the bottle must have to yield a proper base. This is done by reducing blow time until the bottom comes out or some other feature of the bottle is not fully blown. Then reduce exhaust time until either the machine stops on a "blow pressure in mold" or similar error or when you can hear a noise when the blow nozzle lifts. Add 0.1 s to exhaust time to make sure it is always enough.

With these settings you can now see if your oven system is capable of heating preforms up to the right temperature.

On single-stage machines it is all about how fast you can cool down the preform to blowing temperature while blowing parameters can usually be dialed in without affecting cycle time. Most important are hold and cooling time with injection time playing a minor role. Since injection time also controls preform temperature distribution it is best to adjust it with those considerations in mind rather than cycle time.

Most machines do not display the preform temperature and it is up to the operator to manually check by running preforms without blowing them. You should be able to squeeze them but the insides should not stick together. If they crystallize (turning white) they are also too hot.

Another, often less obvious limitation on running faster is the fact that the preform has to come off the injection core without sticking. While this is obvious when the preform totally crumbles it can stick just a bit as shown in Fig. 10.7.

The slightly shorter preform will blow into a different bottle and it takes some observational skill to see this while the machine is blowing bottles. Increasing the draft angle on the injection core can improve on this problem.

In short cycle times on SS machines should follow the best process yielding the highest output of quality bottles. Chasing small increments of cycle time reductions is often counterproductive.

10.6 Troubleshooting of Specific Problems

In this chapter, we will go over the most common processing problems found in RSBM. There is often more than one possible cause—and more than one solution to a particular problem even from the same cause. Not all machines have the features that may be mentioned from time to time,

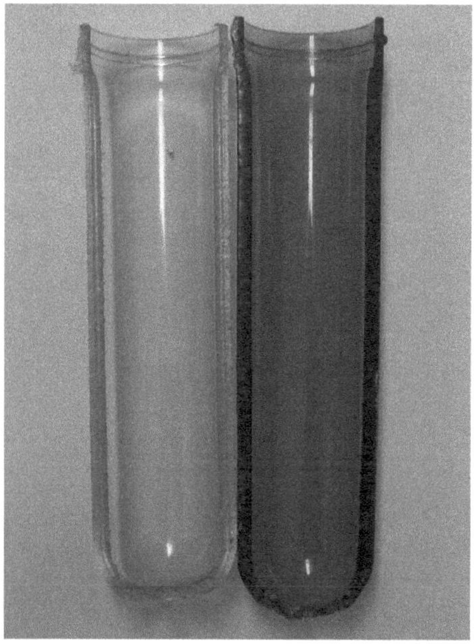

Figure 10.7 Same preform tool but the left one is pulled back by the injection core.

so other solutions may have to be found. Because solutions in single stage are often quite different than in two stage, they are added to the end of the discussion.

Internal Folding in the Neck Area

A number of processors with older machines have this problem, which usually shows up as ring of thick material at the start of the bottle shoulder (Fig. 10.8).

Causes

1. Insufficient heat in the area underneath the NSR.
2. Preblow pressure too late or too low.

Solutions

1. Increase heat in zone #1. If that leads to overheating and haze, increase fan cooling.
2. Move oven bank slightly lower.
3. Push lamp #1 closer to the preform.
4. Mark preforms as described in Section 10.1. Reduce heat in weak areas especially the base. This strengthens these areas of the preform allowing more material to be pulled out of the neck.

Figure 10.8 Material folding in neck area.

5. Reduce preblow pressure delay in combination with
6. Increase preblow pressure: while proceeding in this way, occasionally turn high-pressure off, ensuring that the preblow pressure is not creating too big a bubble.

Single Stage

In many cases this is caused by improper preform design. The counter-intuitive solution is to add material to the area underneath the neck (whereas in two stage we would reduce material there). The thicker material will retain more heat and then stretch out the excess material. Other measures:

1. increasing preblow delay,
2. speed up injection to reduce the temperature difference between gate and neck, and
3. relieve conditioning core in the top area to retain more heat there (when available).

Excessive Material in the Base of the Bottle "Candle Stick"

This defect consists of unsightly accumulations of material in a ring or half-ring shape around the inside center of the bottle. External base folding has the same causes and solutions (Fig. 10.9).

Causes

1. Preblow pressure too late or too low: Material is allowed to gather around the stretch rod, cooling down as a result, and becoming too cold and thick to blow out during high pressure blow.

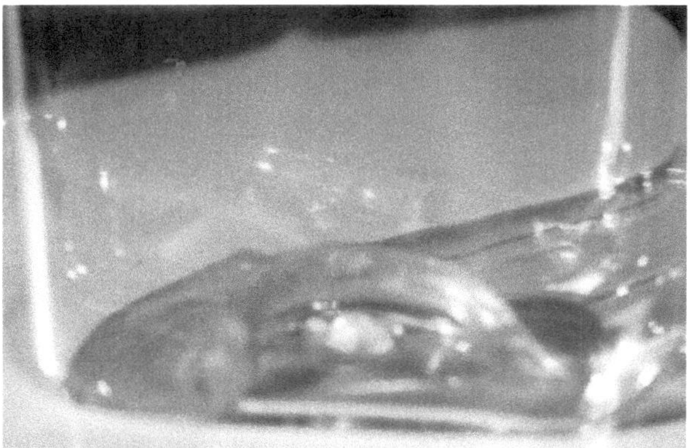

Figure 10.9 Excessive material in base.

2. Preform base too hot.
3. Combination of blow pressure too low and base too hot.

Solutions

1. Increase preblow pressure: While proceeding in this way, occasionally turn high-pressure off, ensuring that the preblow pressure is not creating too big a bubble.
2. Decrease preblow delay: If the delay is already at zero it might indicate that the preblow valve is opening late. Try replacing it.
3. Move the switch indicating the end position of the stretch rod away from the bottle base until the gate goes off-center, then move it back a little. It may be taking too long for the high-pressure air to reach the bottom of the preform.
4. Decrease heat to the base of the preform.
5. Increase blow pressure to a maximum of 40 bar (580 psi).

Single Stage

The issue has similar causes but some of the measures to take are different:

1. Increase injection and/or hold time.
2. Reduce high-pressure delay in increments of 0.02 s. When all gates go off-center add 0.03 s.
3. Change conditioning core to touch the base of the preform (when available).

Off-Center Gate

Whenever the preform gate is not exactly in the center of the bottle base, the wall thickness of the bottle becomes uneven (Fig. 10.10). If for example, the stretch rod tip is skewed to the left, the material on the left will reach the mold wall earlier and more material will harden there even with a perfect temperature profile around the circumference of the preform.

Causes

1. Preblow pressure too high: This pressure can become high enough to blow the preform off the stretch rod. Minute temperature differences around the circumference of the preform drive the preform toward the cooler side.
2. Preblow pressure too early: If preblow pressure commences before the stretch rod is firmly engaged in the preform bottom, the gate may wander off the center.
3. High-pressure air too early: The switch indicating the end position of the stretch rod may be not close enough.

Figure 10.10 Whenever the gate is not in the center of the bottle base, wall thickness is skewed in the same direction as the gate.

4. Stretch rod incorrectly set: Stretch rods should be 1/2–1 mm (0.020–0.040 in.) higher from the base insert than the preform gate wall thickness. As that distance increases, the preform may slip to one side.
5. Stretch rod bent: As neck finishes become smaller, as is often the case for custom containers, stretch rods have to be smaller too. The smaller in diameter they become the easier they bend when they hit a cold preform, for example. This is easy to see: the gate will always be skewed to the same side. Check several bottles and see where the gate is in relation to the recycling symbol or some other engraving. Other causes will push the gate randomly.
6. Preform intrinsic viscosity (IV) too low: When preforms are underdried or overheated during injection their IV may drop significantly and they may blow off the stretch rod.

Solutions

1. reduce preblow pressure,
2. increase preblow pressure delay,
3. move stretch rod switch closer to end of stretch rod or increase blow delay,
4. readjust stretch rod,
5. take stretch rod out and roll over a plane surface. This will show any distortion, and
6. try preforms from a different gaylord or batch. Check IV if necessary.

Single Stage

Causes are similar. Many machines rely on timers to control preblow and blow and setting them can be confusing as there is no clear indication at what time the stretch rod has actually fully extended. When the delay high pressure is too short all gates will be off center and this behavior can be used to find out about the timing.

As single-stage machines make their own preforms proper drying of the resin and control of injection is paramount to guarantee an acceptable (small) IV drop.

Haze in Bottle Walls

Cloudiness or haze first shows when temperature induced crystallinity reaches around 3%. It should not be confused with gate crystallinity, which is always a preform defect whereas haze can be created in both injection and blow molding. Whitish rings or streaks right around the preform gate indicate gate crystallinity whereas haze can occur anywhere on the preform, with prevalence toward the bottom. Haze usually shows as a milky coating on the *outside* of the bottle (Fig. 10.11).

Causes

1. Haziness already present in the preform: It is not unusual to see these defects in preforms.
2. Preform overheats in the blow machine oven: When preform temperature comes close to 120°C (248°F) preforms may crystallize during equalization as they cool down.

Figure 10.11 Overheating the preform leads to haze.

3. Mold temperature may be above 65°C (149°F).
4. If haze happens randomly and is not found in the preforms, preforms with a higher initial temperature may have become mixed with the colder ones for which the process was adjusted.

Solutions

1. check preform supply first
2. reduce lamp settings, increase fan cooling, or speed up machine
3. reduce mold temperature to 60°C (140°F) or less
4. ensure all preforms are at the same initial temperature

Single Stage

It is possible but unlikely to create haze by overheating preform areas with the conditioning heaters (if available). Turning them off for a few cycles will show if they are responsible.

Most likely the preform has not been cooled down enough. Increase hold and/or cooling time until haze disappears.

Pearlescence or Stress Whitening

Also referred to as stress whitening, this defect shows up as whitish rings not unlike pearls, hence the name. They are actually microcracks in the PET molecule structure. They are always on the *inside* of the bottle and show as a milky coating (Fig. 10.12). If there is doubt whether whitening observed in the bottle is haze or pearlescence a simple test can be done: if the affected areas can be scratched off with a finger nail on the inside of the bottle, it is always pearlescence. This can be understood from the knowledge that the inside of the preform has to stretch further and therefore also breaks first.

Causes

1. Preforms are overstretched during blowing. They are either too cold or too thin. The difference can be determined by checking the wall thickness of the affected areas. If they are very thin, the preform may be too hot in this area while other areas are too cold. In this case lamps pointing at the pearlescent part(s) may be at too high a setting. If the affected area is of normal or above normal thickness the preform was too cold before blowing.
2. Preforms are too cold overall.
3. If pearlescence happens randomly preforms with a lower initial temperature may have become mixed with the warmer ones for which the process was adjusted.

Figure 10.12 The defect always shoes in highly stretch areas whereas haze can be anywhere in the bottle.

Solution

1. Reduce relevant lamp settings to areas that are too thin while at the same time increase lamp setting to areas adjacent to the affect areas. This will move material into the overstretch parts of the bottle leading to thicker walls.
2. Increase overall lamp settings if wall distribution is acceptable.
3. Ensure all preforms are at the same initial temperature.
4. Reduce fan cooling and thus increasing oven temperature often does not lead to success as this measure mostly effects the outside preform wall.

Single Stage

Generally the preform is too cold and this can be rectified by decreasing hold and/or cooling time in increments of 0.5 s. If a conditioning core is available, make it touch the areas that are affected: they will not stretch as much and this will reduce pearlescence. Pearlescence is not common in single stage because the inside stays warmer than the outside by the nature of the process.

Deformed Necks

As preforms heat up in the ovens, necks may be subject to an increase in temperature close to the glass transition temperature. PET becomes rubbery and necks deform relatively easy at this temperature. Deformation can show as a crushed NSR, shavings of parts of the neck, or an ovalized or bent neck (Fig. 10.13).

Causes

1. Preforms may have been damaged during transport.
2. Mandrels may be too hot, transferring heat into the preform neck. This becomes more of a problem with wide-mouth bottles.
3. Fan or water cooling for necks may be insufficient.
4. Blow nozzles may put excessive pressure onto the neck.
5. Heat shield may expose preform neck to too much heat.
6. Preforms may not be sitting properly on the mandrels and are then damaged in the blow mold.

Solutions

1. check preform supply first
2. push heat shield closer to preform and/or increase shield cooling
3. increase fan cooling
4. reduce interference between blow nozzle and preform neck
5. check for proper seating of preforms on the mandrels

Figure 10.13 When necks become too warm they may bent during handling of bottle or preform.

Single Stage

This is very uncommon in the process. What can be more of an issue is flashing of the tool that forces the neck inserts open and let material enter this area. Countermeasures are reducing injection speed and/or hold pressure.

Bottle Features not Shaped Properly (Underblown Bottle)

This defect may affect a petaloid base or detail features anywhere in the bottle.

Causes

1. blow mold cooling water too cold
2. high pressure not high enough
3. high pressure too late: "Temporization" time (Chapter 8, Section 8.2) may be too long and the preform may have cooled too much. Or the blow valve is too far away and it takes too long for the blow air to reach the cavity
4. lamp setting of affected bottle area too low
5. air vents may be blocked
6. gate off-center

Solutions

1. Increase blow mold temperature up to a maximum of 60°C (140°F).
2. Increase high-pressure air. If pressure above 40 bar (580 psi) is needed, check bottle design. Radiuses may be too small.
3. Reduce "temporization" time or move stretch rod switch closer to the end of stretch rod.
4. Increase lamp setting of affected area.
5. Check and clear air vents in nonforming areas.
6. See relevant section.

Single Stage

The preform may be too cold or the high pressure starts too late or too low. Decrease hold and/or cooling time for the former and decrease the delay high-pressure blow for the latter cause.

Flats on Bottle Split-Line

Rather than blending with the rest of the bottle circumference, flats above 0.5 mm (0.020 in.) are visible to the eye.

Causes

1. preblow pressure too high
2. blow mold face vents too deep

3. preform outside skin too hot
4. clamp opens during blow
5. blow mold halves not aligned

Solutions

1. reduce preblow pressure
2. check vents, they should not exceed 0.20 mm (0.08 in.) in depth
3. reduce lamp output or increase fan cooling
4. check pressure compensation valve and circuit
5. check alignment and status of leader pins and bushings

Single Stage

Typically, a hydraulic problem with the blow clamp circuit.

Rings Forming in Bottle Body

A ring of unblown material may form in any part of the body. Material did not blow into a single bubble during stretching but developed into two bubbles. The ring then forms where the two bubbles meet (Fig. 10.14).

Figure 10.14 When the preform collapses during preblow the insides touch and form a ring.

Causes

1. Preform collapses because preblow pressure too low or too late.
2. Uneven heat profile makes some preform sections too weak.

Solutions

1. Increase preblow pressure and or decrease preblow delay.
2. Reduce lamp temperature to the center portions of the preform.

Single Stage

To change the heat profile you can change injection speed and/or conditioning cores (if available).

Wall Thickness Over Circumference of Bottle not Uniform

Due to the self-leveling behavior of PET, round bottles will have a circumferential wall thickness distribution varying by 0.1 mm (0.004 in.) or less. Exceeding this range often indicates a problem in the blow process.

See "Gate off center" discussed earlier for this cause.

Causes

1. Rotation of preform not uniform; they may "wobble" in the oven with the result that one side received more heat than the other.
2. Low stretch ratio.
3. Preform wall thickness distribution varies by more than 0.12 mm (0.005 in.).
4. Uneven, plugged, or shallow venting.

Solutions

1. Check oven track for impediments to preform rotation and mandrels for too much clearance to preforms.
2. Total stretch ratio needs to be 8 to achieve the full effect of self-leveling.
3. Check straight seating of preforms on mandrels.
4. Check representative sample of preforms for wall thickness variation.
5. Check vent depth and cleanliness.

Single Stage

Viscous heating is often the main culprit (Chapter 8, Section 8.7). Reducing injection speed sometimes helps.

Excessive Changes in Bottle Volume with Age

Insufficient orientation may lead PET to relax prematurely and shrink during storage. This behavior is also dependent on storage temperature and humidity and many end-users expose bottles to a controlled environment for a varying number of hours after which they measure bottle volume. Volume shrinkage of up to 3% (1.5% for heat-set containers) in 72 h after blowing is acceptable for most applications.

Causes

1. blow pressure too late
2. blow pressure too low
3. temporization too long
4. preform too hot

Solutions

1. Move switch for stretch rod end position away from bottom of mold (linear machine) or adjust high-pressure cam to earlier degree setting (rotary machine).
2. Increase blow pressure.
3. Reduce or eliminate temporization time.
4. Reduce lamp settings or increase fan cooling.

Single Stage

Bottles where this can become an issue are usually not made in this process.

Bottle Fails Burst Test

A burst test is a common procedure to determine whether CSD bottles can withstand the pressures to which they are exposed during filling. Bottles filled with water are exposed to 9–10 bar (130–145 psi) for a fixed time interval and should not fail. Some procedures now require all bottles to be burst and the pressure recorded for statistical analysis. Bottles must only fail in the sidewalls to pass. Lighter bases most often perform better than heavier ones because of higher orientation (Fig. 10.15).

Causes

1. bottle base not oriented sufficiently
2. preblow pressure too late
3. preblow pressure too low
4. temporization too long
5. excessive preform gate crystallinity or voids in preform gate

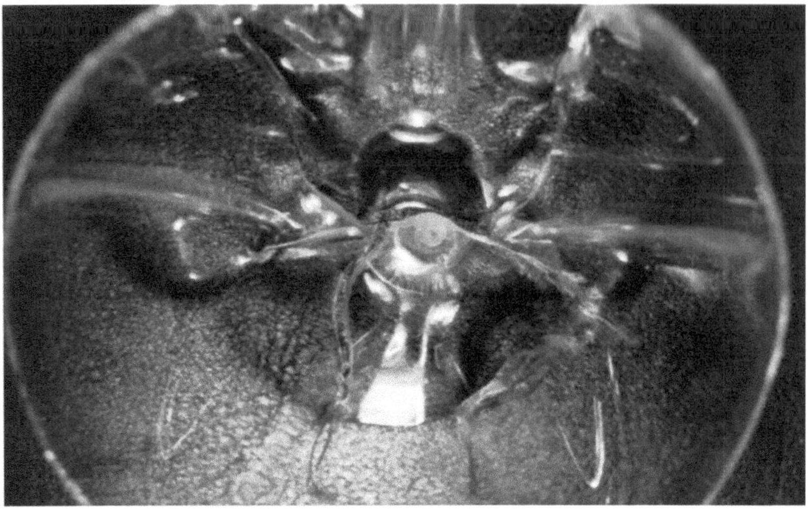

Figure 10.15 Burst bottle base. Bottle base failed in burst test.

6. stretch rod hitting preform gate area too hard causing micro-cracks
7. material viscosity below 0.78
8. high humidity/temperature conditions cause free volume relaxation turning bottles brittle

Solutions

1. increase heat to preform base
2. reduce preblow delay
3. increase preblow pressure
4. reduce or eliminate temporization time
5. check preform supply for crystallinity and gate voids
6. adjust dampening of stretch rod cylinder or move stretch rod away from mold base in increments of 0.1 mm (0.004 in.). Check stretch rod cams on rotary machines for wear on dampening section
7. check viscosity of preforms
8. reduce temperature/humidity exposure of bottles

Single Stage

Bottles where this can become an issue are usually not made in this process.

Uneven Axial Wall Distribution

There are many factors affecting the axial wall thickness of a blown PET bottle.

Not all of them, such as preform design, can be influenced with machine control parameters. Table 10.1 summarizes the effect of the various parameters on the material distribution into shoulder, body, and base of the bottle. It should be noted that this is only a very general outline for standard configurations and that some preform/bottle combinations behave quite differently from the parameters below. Always bear in mind that hotter areas blow thinner.

Single Stage

Axial wall distribution is more difficult to control in this process because it depends so much on the heat profile of the preform that is imparted by the hot runner. Preblow timing is your best bet but you may also try to vary injection speed (Chapter 8, Section 8.4).

Environmental Stress Cracking of CSD Bottles

As mentioned earlier CSD bottles develop 5 bar (70 psi) internally and are therefore more prone to succumb to internal forces and burst, usually at the base. It is often organic liquids that act as stress-cracking agents, after

Table 10.1 Effect of the Various Parameters on the Material Distribution

Machine Parameter	Bottle Shoulder Thickness	Bottle Body Thickness	Bottle Base Thickness
Preblow earlier	+	−	−
Preblow later	−	−	+
Preblow higher pressure	+	+	−
Preblow lower pressure	−	−	+
Stretch rod faster	−	+ or −	+
Stretch rod slower	+	− or +	−
Temporization longer	−	−	+
Temporization shorter	+	+	−
Preform base cooler	−	−	+
Preform base warmer	+	+	−
Preform shoulder cooler	+	−	−
Preform shoulder warmer	−		+
Preform body warmer	+	−	+
Preform body cooler	−	+	−

the blow molding process, but there are a number of measures blow molding plants can take to help prevent this failure. Storage is one of them. If bottles are exposed to high temperature and humidity conditions, a process called free volume relaxation sets in. The space between molecular chains that was locked in during rapid cooling decreases and the chains move closer together. As a result, yield strength increases as the chains can no longer slide beside each other but the material becomes brittle with sudden failure under stress (Fig. 10.16).

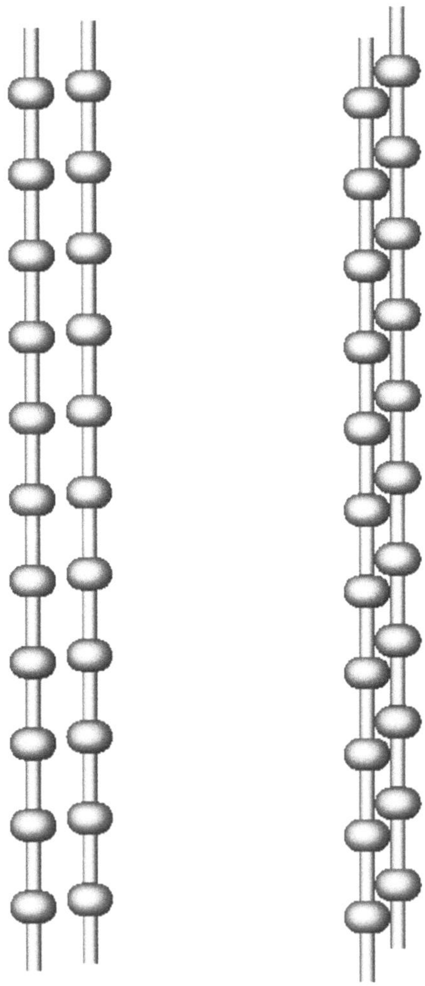

Figure 10.16 During free volume relaxation, molecular chains move closer together causing a change from ductile to brittle behavior.

Causes

1. Unsuitable material distribution in base.
2. IV drop in preforms due to improper drying.
3. IV drop in preforms due to high usage of regrind.
4. Bottles are exposed to alkali line lubricants or detergents.
5. Bottles are exposed to high storage temperatures and/or humidity.

Solutions

1. Change heat profile and/or blow parameters.
2. Check preforms for IV degradation. Minimum IV for CSD is 0.78 in the preform.
3. Check with preform supplier whether regrind is used and in what percentages (maximum 10%).
4. Ensure only nonalkali line lubricants and detergents are used.
5. Maintain good air circulation in warehouse and use first in first out (FIFO) stock rotation.

Single Stage

Bottles where this can become an issue are usually not made in this process.

Cracked Gates

The preform gate has to withstand the substantial impact caused by the stretch rod hitting it at considerable speed (Fig. 10.17).

Causes

1. Preform bottom too cold.
2. Stretch rod speed too high in combination with thin stretch rod tip.
3. Gate area of preform not properly formed during injection molding.

Solutions

1. Increase heat to preform bottom in small amounts.
2. Decrease stretch rod speed. In rotary machines use a slower stretch rod cam.
3. Check preforms under polarized light to see if gate area has formed properly (Fig. 10.18). (Note: A customer in Colombia has discovered that using glasses handed out in 3D movies can be used in lieu of a polarized light table. Put

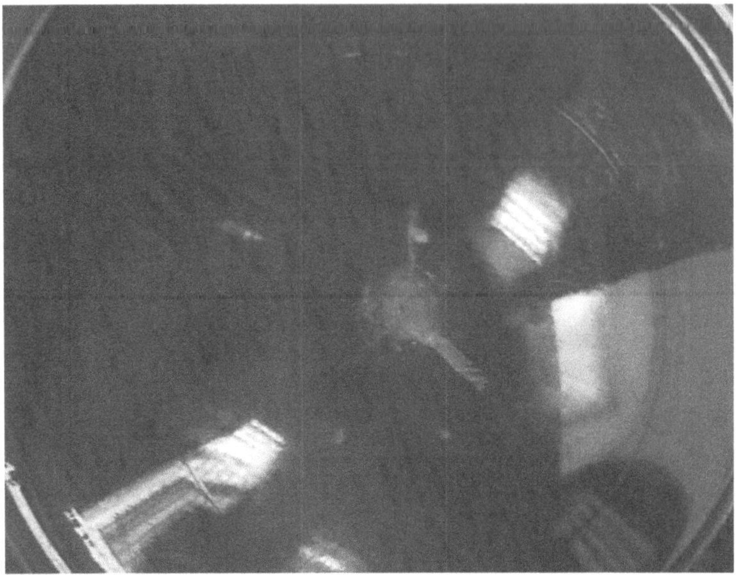

Figure 10.17 Gates may crack when hit by the stretch rod.

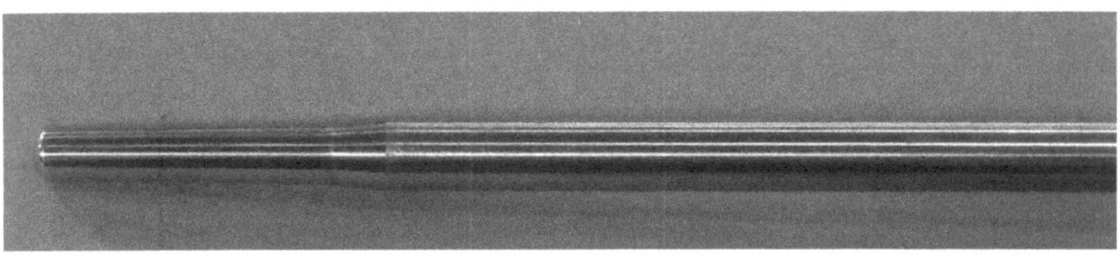

Figure 10.18 Stretch rods are thinned out to allow better stretching of the preform base but the higher pressure they exert on the gate may lead to cracking.

the glasses on and hold the preform against a light computer or phone screen and you will see the lines the same way as on a polarized light table!)

Single Stage

This defect is usually not found in this process as the gate is too warm to crack.

Drop Impact Failure

Properly blown PET bottles typically do not fail a drop from 1.50 m (5 ft.) that is often required for qualification. However, if certain process conditions come together, this may happen and is more common in single stage.

Causes

1. insufficient orientation (strain hardening)
2. crystallization in the preform, leading crystallized areas to become brittle
3. sharp transition between thick and thin sections
4. Pearlescence (see earlier)

Solutions

1. Run a cooler preform by turning lamps down/increasing ventilation or reducing hold and/or cooling time in single stage.
2. Check preforms for crystallization and avoid it in single stage by cooling it more.
3. Even out wall thickness between thick and thin sections.

Top Load Test Failure

As bottles are being ever more lightweighted top load becomes more of an issue as it is almost exclusively dependent on wall thickness (Section 10.4). Bottles will always fail where they are structurally the weakest and the operator needs to know where the failure occurred to direct more material into the failed area.

Causes

1. uneven wall distribution in the circumference
2. weak heel
3. weak shoulder

Solutions

1. see earlier
2. increase preblow delay to force more material into the heel
3. decrease preblow delay to force more material in the shoulder

Single Stage

The same solutions apply; in addition, changing the injection speed will also redistribute material into areas that need it.

Panel Sink

All plastic bottles will end up with depressed panels over long periods of time (6–18 months). This is because of moisture loss that creates a vacuum inside the bottle. The defect described here is different in that it can be seen and measured right after blowing or while empty bottles are

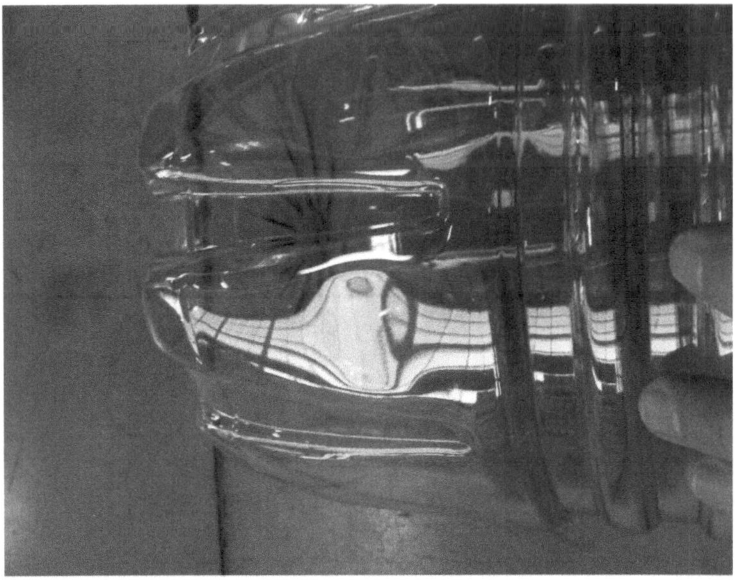

Figure 10.19 While more common with oval bottles, panel sink can also happen with round ones.

in storage. It is more common with oval bottles but can also happen with round ones. This creates problems when bottles are printed or labeled in the empty state. In most cases, the bottle wall is not strong enough to sustain its shape (Fig. 10.19).

Causes (1)

1. preblow pressure too high or too early
2. wall thickness too low
3. mold too cold
4. mold walls not convex enough
5. venting insufficient

Solutions (1)

1. Lower preblow as low as 5 bar (90 psi) and run a cool preform. Do a preblow audit to make sure the preblow pressure does not push the material against the mold walls.
2. Make sure stretch rod is centered and wall thickness on both sides of the panel area is even.
3. Raise mold temperature to a maximum of 45°C (113°F).
4. Clean vent channels; add hole vents in corners.
5. Check mold walls for convex shape; all mold walls should be slightly convex so the bottle walls can shrink to the correct shape after blowing.

Cause (2)

The center panel is too thick and cools (shrinks) more than the thinner side parts. The thick center panel is sometimes referred to as a "spine."

Solution (2)

This often happens with oval bottles when the preform diameter and the short side of the bottle are nearly identical. Minimal preblow pressure (just enough to avoid collapse of the preform) and high mold temperature combined with extra venting are the tools to use.

10.7 Defects Particular to Single-Stage Molding

All of the defects listed in Section 10.6 may also happen in single-stage molding. However, an additional set of defects may be present as a result of problems in the injection part of the machines but then show up as bottle defects. Here is a selection from my experience.

Bottle Cloudiness

Haze or cloudiness is either caused by insufficient cooling of the preform or problems with the dryer (Fig. 10.20).

Causes

1. insufficient cooling time
2. insufficient hold time
3. insufficient drying

Figure 10.20 Cloudiness or haze is always created during injection.

Solutions

1. increase cooling time (compare with values in Chapter 8, Section 8.3)
2. increase hold time (compare with values in Chapter 8, Section 8.3)
3. check for dryer alarms and compare temperature/residence time to values in Chapter 3, Section 3.3

Bubbles in Preform/Bottle

Bubbles may appear randomly in both time and space and can be caused by air or water inclusions in the melt (Fig. 10.21).

Causes

1. Insufficient back pressure prevents air surrounding pellets from being squeezed out.
2. Barrel temperature in feed zone too low.
3. Insufficient drying time or temperature.
4. Water leakage somewhere on the extruder or from pipes on the ceiling.

Solutions

1. Increase back pressure; if over 20 bar (290 psi) is necessary to achieve the desired result, screw may be worn.
2. Increase barrel temperature #1 up to 285°C (545°F).
3. Compare temperature/residence time to values in Chapter 2, Section 2.3.
4. Check for water leakage; a possible source would be the extruder water-cooling jacket.

Preform Stringing

Incomplete separation of the hot melt from the cold preform vestige resulting in fine thread protruding from the preform. Usually more frequent in high temperature/humidity environment (Fig. 10.22).

Causes

1. temperature too high
2. temperature too low
3. insulator worn

Solutions

1. Reduce nozzle temperature. If there are no individual nozzle temperature controllers reduce overall hot runner temperature.

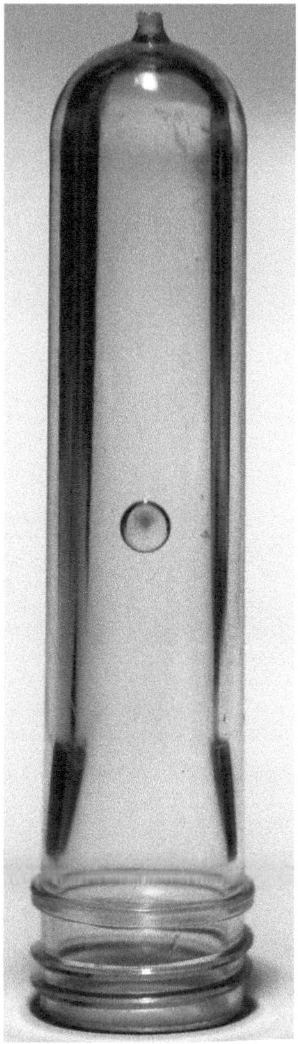

Figure 10.21 Bubbles may be caused by air or water inclusion.

2. Increase nozzle temperature. If there are no individual nozzle temperature controllers increase overall hot runner temperature. This may sound not very scientific but it is sometimes difficult to predict how the material reacts to temperature changes.
3. Replace worn insulators.

Figure 10.22 Stringing usually happens in the hot summer months in nonair-conditioned machines.

Sink Marks

Voids can appear anywhere on the preform but are usually found near the neck area. Pressure was not high enough to fill the preform completely or the screw bottomed out on the barrel.

Causes

1. insufficient hold pressure/time
2. insufficient injection pressure
3. insufficient injection cushion

Solutions

1. Increase hold pressure/time according to values given in Chapter 9.
2. Increase injection speed to get higher injection pressure but stay below 100 bar (1450 psi).
3. Increase shot size so the injection cushion is at least 3 mm.

Gate or Body Splay

A yellow to brown discoloration affects gate or body of the preform. It is caused by excessive shear stress (Fig. 10.23).

Causes

1. injection speed too high
2. nozzle temperature too high
3. insufficient decompression of the hot runner

Solutions

1. reduce injection speed
2. reduce nozzle temperature or entire hot runner if there are no individual nozzle temperature controls
3. move screw back after hold and before shut-off nozzle closes

Unmelts

Unmelted or partially melted resin pellets show up in preforms and bottles.

Causes

1. insufficient drying time and/or temperature
2. barrel temperatures too low

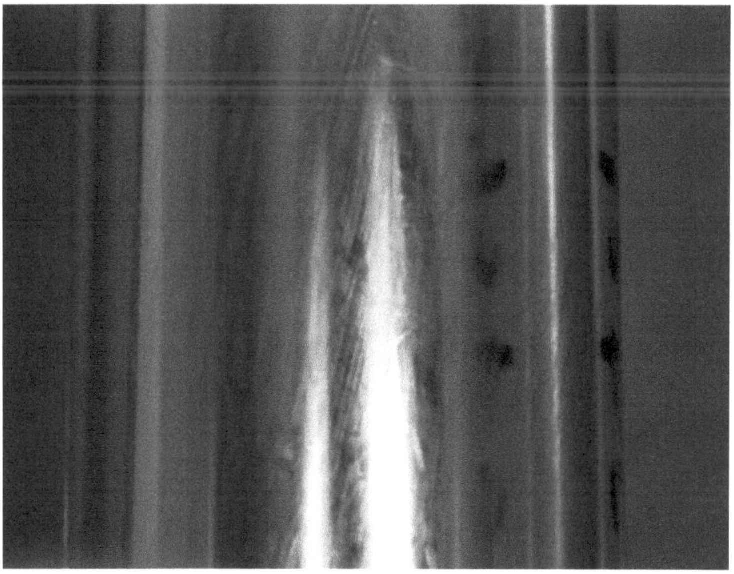

Figure 10.23 Splay can take on different shapes.

3. resin inlet temperature too low
4. extruder back pressure too low

Solutions

1. Compare temperature/residence time to values in Chapter 2, Section 2.3.
2. Increase barrel temperatures to a maximum of 295°C (563°F).
3. Resin inlet temperature should be around 160°C (320°F).
4. Increase back pressure to 15–20 bar (200–300 psi).

Short Shots

The injection cavity is empty (except for air) when injection begins. At the end of hold time the preform must be fully formed (Fig. 10.24). When not enough melt is injected parts of the neck finish may not form, a problem most people call a short shot. Here are the reasons why this may happen.

Causes

1. The shot size (the screw position at the end of recovery) is not set correctly.
2. Injection time is too long (injection rate is too slow).
3. Hold pressure is too low.
4. Hold time is too short.
5. If the tool has valve stems they may not be open completely during injection or open too late because they are sticky.
6. Material is too cold.

Solutions

1. set the shot size so there is a "cushion" of plastic of at least 3–5 mm when the screw has reached the end position
2. increase injection speed

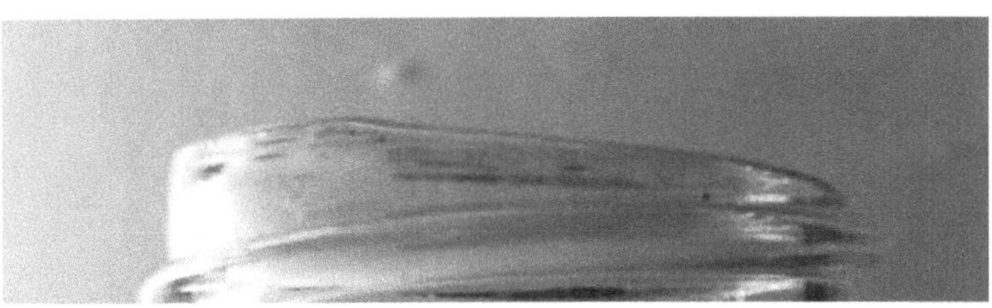

Figure 10.24 Short shots happen when there is not enough material or pressure to fill the cavity.

3. increase hold pressure 10 bar (150:psi) at a time
4. increase hold time 0.5 s at a time
5. air pressure for the valve gates should be 10 bar (145 psi). With thermo-gated hot runners check for proper insulators
6. increase barrel and hot runner temperature

10.8 Summary of Preform Quality Checks

From the previous sections it is obvious that the blow molding process does not cause all bottle defects. Here are the main criteria for the inspection of incoming preforms, whether they are purchased or made in-house.

- wall thickness distribution
- weight
- physical dimensions and perpendicularity
- thread integrity
- gate crystallinity
- contamination
- bubbles
- haze
- if the blow machine does not automatically reject deformed preform necks, these should be visually inspected at random intervals. This will not eliminate the one crushed neck but may identify a gaylord full of bad preforms

AA content can be checked in the preform or the blown bottle. Most of the AA present in a bottle was actually created during the injection molding of the preforms. Checking preforms before making bottles is often advisable and avoids producing a batch of reject bottles and then having to scrap them.

11 Economics

Chapter Outline

Ever-increasing outputs in both injection and stretch blow molding have led to preforms and bottles becoming something of a commodity. Converters charge for conversion plus the price of polyethylene terephthalate (PET) at current rates for standard water and carbonated soft drinks (CSD) bottles. Other, lower volume containers still fetch a premium but the time will surely come when many more bottles have become standard and the premium will vanish. Even multilayer preforms are starting to become

Stretch Blow Molding. http://dx.doi.org/10.1016/B978-0-323-46177-1.00011-1

Figure 11.1 PET lends itself to a variety of bottle shapes and applications. Photo courtesy of Siapi.

widely available, entering the competitive race (Fig. 11.1). In order to understand the economics behind the pricing of PET bottles, here is a look at what is driving prices, how costs can be cut, and what should be considered before embarking on a new product.

11.1 Container Types

CSD Bottles

As was mentioned in the Introduction, CSD bottles were the first big hit in PET's success story. Initially, bottles were blown with a hemispherical base and a base cup was glued onto it to allow it to stand. Then a one-piece bottle was invented and today the overwhelming majority of these bottles uses a five-footed (petaloid) base and a PCO, Alcoa, or BPF 28-mm neck finish (Fig. 11.2).

Despite several attempts to change the base design, the petaloid base has survived from the early days of one-piece bottles because of its performance and moldability. Many improvements have been made to perfect the design and there are active patents protecting their ownership. Potential end users may have to pay a license fee to use the latest designs.

Even not-so-metric North America has standardized on the 28-mm neck and actually all neck dimensions are metric anywhere in the world. The inner diameter of the 28-mm neck is wide enough to allow the stretch rod to pass through, even considering the mandrel wall thickness in between. It is also small enough to withstand considerable hoop stress at a thickness of approximately 2 mm. Grooves or interrupted threads allow carbonation to escape safely when the consumer opens the bottle.

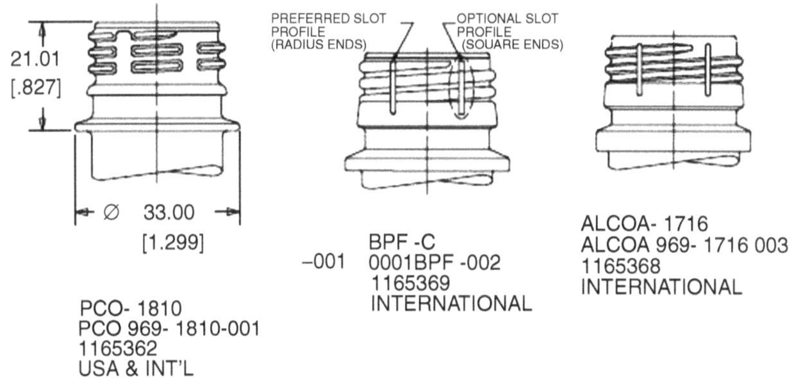

Figure 11.2 Most common neck finishes for CSD bottles.

Because of concerns about obesity, industrialized countries have seen a slow decline in CSD bottle sales. CSD bottles are filled with a carbonation of approximately 3.5–4 volumes. Shelf life is determined as the point when this level drops by more than a certain percentage. (There are different standards of maximum carbonation loss between companies and countries; 15% might be an average value.) Volume-to-surface ratio, gram weight, material properties, degree of orientation as well as temperature and humidity, all play a role in how quickly this level will be reached. The volume-to-surface ratio falls as bottles become smaller, that is, in smaller bottles there is more surface area per volume to allow escape of carbonation. Twenty ounce or 500 mL CSD containers will lose 20% of their carbonation within 6 weeks at 30°C. This is still acceptable, as thirsty consumers will have consumed the contents by that time. However, this loss will be significantly faster for packages smaller than ½ L and only multilayer or coated bottles will be able to provide acceptable shelf life. If and when soft drink companies decide to pack CSD in bottles the same size as cans (355 mL or 12 oz) converters will have an opportunity to sell premium packages in large quantities. Significant investments have been made at the time at some point, but to date (early 2016) there are only few CSD packages smaller than 500 mL in the marketplace. One reason may be that large brand owners want to keep their ability to switch between PET, steel, aluminum, and carton depending on the price developments in the underlying resource market.

Water Bottles

Water has been growing very strongly since the early 1990s, contributing over 50% to PET's overall growth. Water bottles are about 30–50%

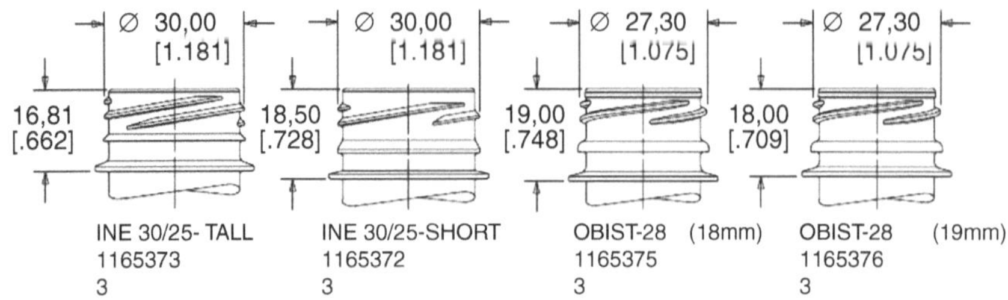

Figure 11.3 Most common water bottle neck finishes.

lighter than CSD bottles because they do not have to withstand carbonation pressure and maintain carbonation levels (although some soft drinks companies that are also in the water business have used existing tooling and produced water bottles with the same features as CSD bottles). Besides the lower weight, water bottles also use different neck finishes (Fig. 11.3).

These necks use three-start threads, instead of one, making it easier for consumers to screw the cap back on the bottle. They are only slightly thinner than CSD necks—principally because it is difficult to force material into smaller crevices during the injection molding process. Lightweight water bottles feature stiffening ribs in the base, allowing less material to be used there.

Heat-Set Bottles

Most juices and "New Age" drinks require the product to be sterilized and filled at temperatures of 80–95°C (Chapter 9, Section 9.4). These bottles have wider necks (between 38 and 43 mm) which is largely due to a historic development in the juice industry and not driven by technical considerations. The premium for these bottles is a result of higher production costs and the higher technical demands of a process that not every converter can meet.

Custom Containers

There is growing demand for noncommodity containers such as personal care, household, and food other than water, CSD, or juice. Necks of custom containers are often smaller than the standard 28-mm neck (unless they are widemouth). Standards for the threads of bottles and closures are maintained by the Plastic Bottle Institute, a division of the Society of the Plastics Industry. Specification of a 24-mm neck of this type with one thread turn is: M24SP400 (…410 would indicate 1.5 turns, …415 stands for 2 turns) (Fig. 11.4).

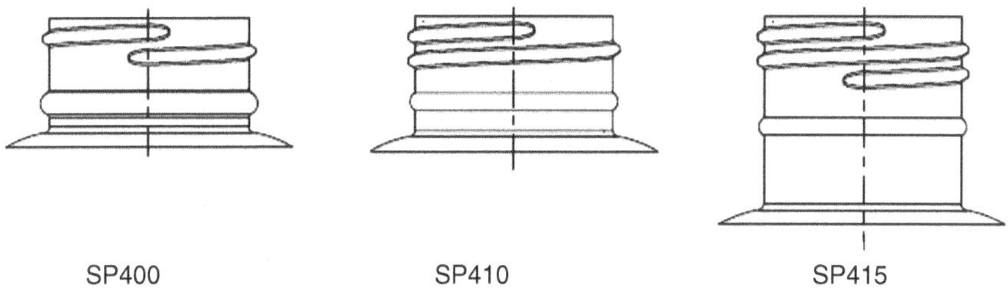

Figure 11.4 Some typical neck finishes of custom containers.

Many custom containers are being manufactured in the single-stage process. This process has several advantages over the reheat stretch blow molding (RSBM) process for this type of bottle:

- Bottles do not require a neck support ring because they are held in the neck insert. Most RSBM machines need special devices to transport preforms without neck support ring.
- Bottles can be packed without any scratches. By contrast, in the two-stage process, at the end of the injection molding cycle preforms fall onto conveyor belts and scratches are unavoidable.
- The start of the neck thread is automatically oriented to the shape of the bottle. This allows the orientation of a rectangular cap an oblong bottle, adding to consumer appeal. In RSBM machines special devices had to be developed for rotating preforms inside the mold and detecting the thread starts with photocells, incurring a cycle time penalty in the process.
- Especially single-stage machines with conditioning stations allow preferential temperature conditioning at certain points on the preform for oval or rectangular bottles.

However, single-stage machines have inherent problems (Chapter 8) that may lead to process inconsistencies. Due to the inefficient use of the blow section, changing bottle types can become quite expensive. RSBM processors have successfully overcome the challenges presented by these bottles and offer competitive packages. The flexibility and quick changeover time of RSBM machines allows processors to react quickly to market demands. A key to success is to match as many bottles as possible with one particular preform (Chapter 11, Section 11.3). If preforms vary only by the number of neck thread turns, it is often possible simply to exchange the neck inserts instead of the complete preform tool.

Beer Bottles

No other package has excited the PET community more over the years than the prospect of PET entering the multibillion dollar beer market. The challenges are numerous as most brewers expect a shelf life of 6 months. Beer is extremely sensitive to oxygen ingress with a maximum acceptable level being 1 ppm (part per million) over the shelf life of the bottle. Beer bottles are also under carbonation pressure, which contributes to the possible oxygen ingress by as much as 15%. Carbonation loss itself must not exceed 10%. Most beer bottles are 500 and 330 mL packages with high surface-to-volume ratios, exacerbating these challenges. Acetaldehyde (AA) migration from the PET wall (Chapter 2, Section 2.6) must be limited to 20 ppb (parts per billion), which is the threshold of detection by a consumer.

The list of solutions that companies have employed reads like a "Who's Who" of material, barrier, and coating technologies and need not be repeated here (Chapter 9, Section 9.8). Taking environmental conditions into account is of great importance. Beer bottles that work well in Central Europe or the North-Eastern United States might fail in Brazil or the Philippines. It follows that there will be more than one solution for PET beer bottles.

PET, and especially polyethylene naphthalate (PEN), beer bottles are still more expensive than comparable glass bottles and many brewers are in no hurry to go through expensive factory retrofitting to accommodate PET bottles and still pay more for the package. However, PET's advantages of being shatter-proof and lightweight will eventually lead to the construction of high-cavitation tools for either multilayer or coatings that may bridge this gap. Consumer acceptance is another unknown and brewers will have to invest in marketing their PET bottles to get results. On a positive note, professional taste testers at Arthur D. Little, Inc. found only minimal differences between glass and beer bottles.

The market share of PET/PEN beer bottles is still minuscule at the time of this writing. It is estimated to be 6 billion bottles, a mere 2% of the 294 billion beer bottles sold worldwide[a] and it is unclear whether beer applications will become another PET success story.

[a]PIRA International, *The Future of Beer in PET Packaging*—a multiclient study for brewers, brand owners, packaging converters, and suppliers, 2011.

11.2 Business Models

There are three ways in which PET bottles are manufactured:

1. Large converters make both preforms and bottles. Bottles are expensive to ship whereas preforms are not. Therefore blow molding plants are mostly within a maximum radius of 500 km (300 miles) of filling plants. These plants may also manufacture their own preforms but more commonly preforms are injection-molded in central plants and shipped to the blow molding facilities. The latter approach has the advantage that injection molding experience need only be spread to a few plants. This model seems to give multilayer bottles a slight advantage over coating solutions, as they can follow the same model, whereas coating technologies and their specific process challenges will have to be introduced into every blow molding plant. Preform plants often have a low-cavity blow molder in-house to test new bottle or preform designs. Converters may also be part of a so-called "through the wall" manufacturing model where the blow molding plant is right next to the filling plant and bottles are often literally run on conveyors through the walls of the two plants.
2. Preforms and bottles are manufactured by different companies. There are a number of companies that have specialized in injection molding of PET preforms. They apply large-scale economics with plants operating 365 days/year and 24 h/day. The resulting low price of PET preforms has been a boon for smaller converters that prefer to buy preforms and blow bottles close to their customers.
3. The filling company manufactures bottles and/or preforms either in the filling plant site or using a distributed model. As the technology has matured many end users have taken production in-house to cut costs, especially in transportation, and become independent of suppliers. This is especially true for water bottling companies. Competition in this market is fierce. A filled, capped, and labeled bottle as part of a multipack may retail for less than US$ 0.10. That leaves little room for middle-men when distribution and marketing costs are factored in.

11.3 Tooling Costs

Regardless of the business model used tooling costs is one area where all manufacturers can reap large savings for their businesses. Proper selection of tooling components matching the range of bottles produced and matching new bottle developments to existing tooling are key for keeping tooling costs under control. There are several ways in which this can be accomplished.

Since CSD bottles must be stronger than water bottles (to withstand the interior pressure caused by carbonation) converters routinely use the same preforms for 1 L CSD/1.5 L water or 1.5 L CSD/2 L water. Stretch ratios will be optimum for the CSD bottles and the process adjusted for the larger water bottles. Or companies may use the same bottle for CSD or water even if the water bottle will then be overengineered.

There is not much that can be done on the blow mold side since every bottle needs its own, unique mold. Forcing constant innovation on to mold makers and driving prices for blow molds down in the process has certainly happened and benefited the industry as a whole.

But there is a lot more possible when it comes to preform tooling. Standardization of preform tooling is referred to as "rationalization" and is a powerful concept in today's competitive market place.

For once, companies can standardize on one neck size for their product categories of CSD, water, and hot fill. (Custom containers are more difficult in this aspect since aesthetic considerations and cap selection play such a large role.) Neck inserts are the most expensive part of the injection tooling so that this measure alone can save considerable amounts of money. Next, experienced engineers may review existing preform tooling and find synergies between different preform/bottle combinations. Here are the parameters that need to be studied:

- axial stretch ratio
- internal hoop stretch ratio
- Blow ratio: There are limits to how much PET can stretch. For most grades of PET this limit is 16 times, and PET film (known under the trade name "Mylar") is indeed stretched 4 times in both directions. In bottle blowing, blow ratios above 12 become quite difficult but not impossible to process. Practical limit seems to be about 14.5. This value is also material dependent. Higher IV materials have lower natural stretch ratios (Chapter 2) but are used for CSD bottles with high blow ratios, yielding a high degree of orientation
- Preform weight

- Preform body and gate wall thickness: The gate wall thickness is always smaller than the body wall thickness (except for lightweight water bottles), the ratio between the two being just below 80%. The gate will stay largely unstretched in the bottle and its thickness will determine how much heat it can retain. That in turn, will have an effect on how much it will pull the preform body during stretching
- Preform temperatures before entering the blow molds. If no thermal camera is available (they are still very expensive) the infrared thermal sensor on the machine can be set to different heights and the temperatures recorded. These data can be used to explore how much this temperature may be varied to suit bottles with different stretch ratios
- After section-weighing of bottles the preblow timer and preblow pressure may be adjusted to move material into the bottle shoulder, for example
- Machine setup sheets; this will determine whether a particular machine is suited for specific bottles. Older machines are often limited with respect to heating power, which may be a restricting factor

After analysis, tests are then run to confirm that the preform in question can indeed be used for a particular bottle. The experience gained will then also have to go into new bottle development with the aim of assuring that no new preform tooling need be purchased.

One water bottler reported at a conference a reduction of neck finishes used from 7 to 3, and preforms used from 21 to 12, while increasing the bottle types from 37 to 61 over the same 3 years.

These results not only reduce the cost of tooling but also lead to higher uptime in injection molding machinery. Fewer changeovers reduce the need for expert intervention in the process and generally also create a better-running plant.

Tooling costs play a more significant role today than they did 10 or 15 years ago. Tool prices have actually come down over this time period because of increased competition and more economical manufacturing methods. However, cycle times have dramatically decreased but the total number of shots a preform injection tool can perform before it is worn out, have only marginally increased. This number stands at 10–12 million cycles for horizontal machines and only if proper maintenance is performed at regular intervals. Especially brand owners for water bottles that run the same tool 350 days/year at a cycle time of around 7 s find themselves in the undesirable position to have to refurbish or buy a tool every 2 years!

Companies now plan weight reductions in sync with the forecasted tool changes thereby getting a reasonable payback for the capital outlays.

11.4 Lightweighing of Bottles and Caps

The 50% surge in resin prices in 2008 has further spurned a rush of developments to reduce the weight of bottles and caps. The results have been astonishing:

- 500 mL water bottles as low as 8.5 g from a previous average of 16 g
- 20 oz CSD bottles as low as 19 g from a previous average of 26 g
- Water bottles caps from 2.5 g to less than 1 g

These improvements came from several, unrelated areas.

- Lighter neck designs for both water and CSD bottles.
- Different bottle designs allowing smaller wall thicknesses.
- Improved preform designs.
- Machines built specifically for small, lightweight containers.

Despite a great price drop in PET resin in 2015 to as low as US$1000 per ton light-weighing has continued concentrating now more on bottle shapes and top load and shelf life requirements. One gram of weight-saving for 100 million bottles saves 100 tons of PET and even at the lowest prices we have seen is still a saving of US$100,000. It also saves 80,000 kWh of energy in preform molding, 36 tons of CO_2, leading to further savings. The reduced carbon footprint is good for both the planet and the brand owner's image.

Lighter Necks

This was the first step in the lightweighing cycle. The most common necks for water and CSD bottles were the 30/25 and PCO 1810, respectively (Figs. 11.5 and 11.6). The strategy for both applications was slightly different. Let us look at water first.

- Diameter was reduced from 30 to 25.84 mm.
- Height was reduced from 16.81 to 12 mm.
- Wall thickness was reduced from 1.4 to 1.05 to 1.2 mm.

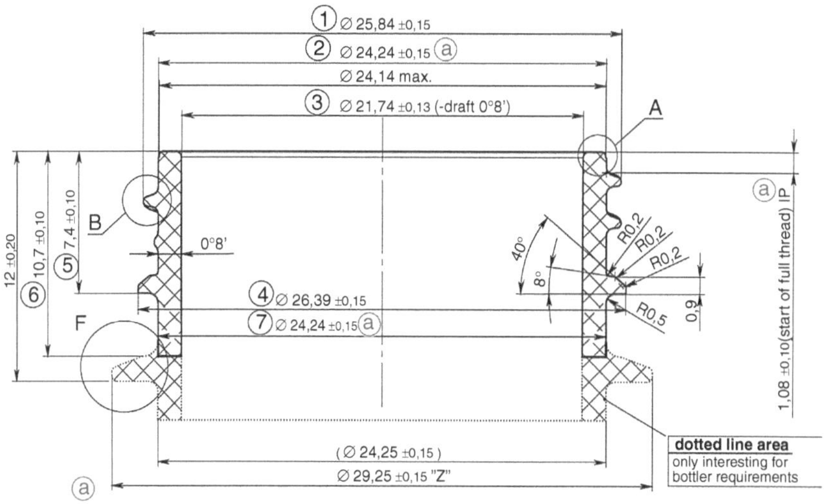

Figure 11.5 Thinner walls and a shorter height contribute to the weight savings of this bottle neck for a 26-mm closure. Drawing courtesy of Bericap.

This resulted in a weight reduction of the neck finish from around 4 g (as measured to the bottom of the transfer ring) to about 2 g. These numbers differ slightly between manufacturers but nevertheless show an approximate and astounding 50% weight reduction. It is little wonder that many companies jumped on this opportunity to save substantial amounts of money.

Another neck has since made considerable inroads. It is 29 mm but weighs only 0.1 g more than the 26 mm. Consumers prefer it as it is easier to pour or drink from.

Similar efforts were made with CSD necks and two competing designs have evolved (Fig. 11.7).

Lighter Caps

The new necks require new caps and designers have used the opportunity to go beyond just fitting a cap on a new neck (Fig. 11.8). Instead, they have pushed the limits of what can be injection molded and have engineered caps that are truly as light as can be, especially for water. The new water cap features a wall thickness as low as 0.95 mm and weighs just under 1 g, down from 1.85 g of the tall 30/25 cap (Figs. 11.9 and 11.10).

The requirements for CSD caps are significantly higher as the caps have to withstand the significant internal pressure of the carbonated contents. Nevertheless, the new CSD cap saves 0.6 g, reducing the weight from 2.95 to 2.3 g.

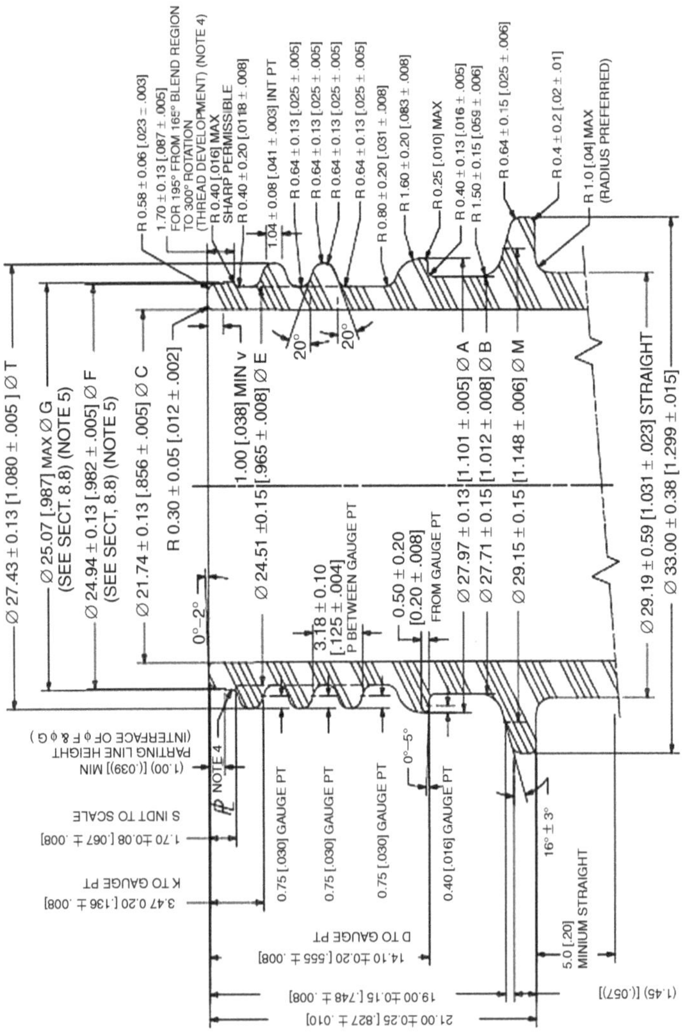

Figure 11.6 PCO 1810.

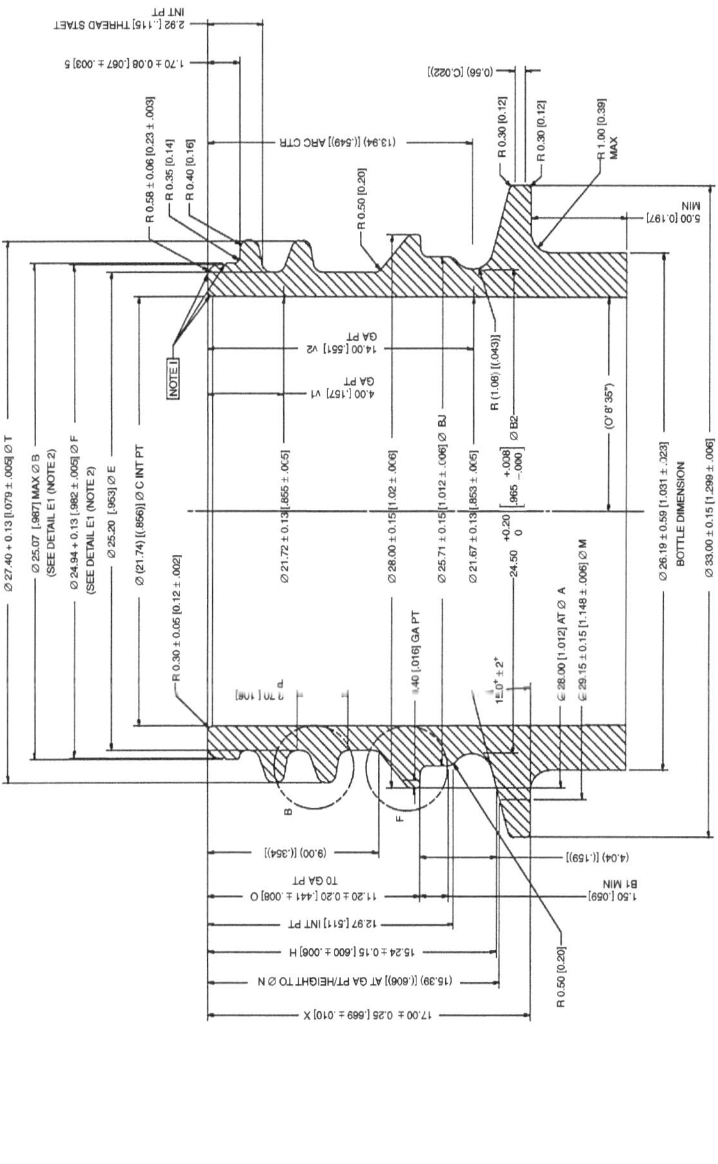

Figure 11.7 PCO 1881. The industry has now standardized on the PCO 1881 and more and more bottles can be found with this neck on your supermarket shelves.

Figure 11.8 Twenty six millimeter caps for water bottles weigh just 1 g. Picture courtesy of Bericap.

Figure 11.9 The 29-mm closure offers more consumer convenience at almost the same weight as the 26.7-mm closure. Courtesy of Corvaglia Mould AG.

Figure 11.10 PCO 1881 closure. Picture courtesy of Corvaglia Mould AG.

New Bottle Designs

The water market is divided into a premium sector that is sold in single bottles at convenience stores and such and the bulk purchase of 12, 24, or 30 bottles sold in supermarkets. Bottle weights in the former were until recently as high as 27 g for a 500 mL/20 oz bottle while the respective weight in the latter was around 16 g. Water offers a great many opportunities to lightweigh because the functional requirements are relatively small.

Top load is the most critical parameter in this respect. It has two components:

- Static load that the bottom layer of a pallet of filled bottles may experience in a warehouse over longer periods of time.
- Dynamic load during transport where loads may bounce and the bottom layer may experience much higher loads for short periods of time.

The top load value is attained by compressing a bottle at a determined speed (usually 1/2 in./m) with an apparatus that contains a load cell. Typical bottle behavior is to slowly decrease in height and then to suddenly buckle. The highest load is measured at the point of buckling and is then recorded as the top load value. Other tests involve loading the container with a certain load (maybe one that compresses it by 10 mm) and leave the load on for a fixed amount of time to see if the material creeps and the container eventually fails. In most cases top load is being measured on empty bottles as part of routine quality control. It has long been this author's contention that results thus obtained do not have validity for fillers whose main concern is top load performance of filled bottles. Here are some examples of top load test conducted at a converter for vegetable oil bottles (Table 11.1).

A more scientific and practical approach is to use finite element analysis before a container is manufactured to estimate where and at what load the container will fail. This assists in container design. Once there are containers that can be filled and the encasing (whether carton box or shrink-wrap) is determined a test load should be equipped with load cells placed at the bottom layer. These load cells can feed data to battery-operated recorders at predetermined intervals and thus give precise records what happened during transport. Static load can easily be calculated once pallet configuration has been established and a pallet layer is then subjected to the calculated load over a period of time in controlled conditions. Values from both sets of experiments can then be used to determine the minimum top load

Table 11.1 Top Load of Empty and Bottles in Kilogram for Various Sizes

Bottle (cm^3)	Empty	Empty	Empty	Average	Full	Full	Full	Average	Increase (%)
200	9.55	10.65	11.65	10.6	36.55	49.2	49.75	45.2	425
250	8.75	8.2	9.15	8.7	38.75	32.75	31	34.2	393
400	6.5	6.3	6.6	6.5	30	43	41.5	38.2	590
500	10.1	9	9.8	9.6	25.75	41.25	36.2	34.4	357
750	17	16.5	16.3	16.6	68	63.5	69	66.8	403
900	11.95	10.95	10.15	11.0	30	35	41.85	35.6	323
1000	13.4	13.2	14.05	13.6	27	28.3	33.9	29.7	219
2000	11.55	12.2	11.8	11.9	34.7	34.65	34.9	34.8	293

Full bottles show large improvements.

required and give designers the knowledge how much further lightweighing can go (Fig. 11.11).[b]

The lightweighing of water bottles has shifted the attention of the industry to this situation and many top load values are now listed for filled bottles.

Bottle designers sought ways of adding mechanical strength to bottle shapes that did not require thick walls and came up with a number of different solutions such as vertical ribs, swirls, and other features. Round bottles for the two markets mentioned earlier weigh now around 16 g for the premium market and 8.5–12.5 g for the take-home market. Oblong shapes require more material and are only used in the premium market (Fig. 11.12).

Improved Preform Design

The lighter bottles also need suitable preforms that still have to conform to all the restrictions of the injection molding process. Foremost, the preform wall thickness has to be adapted. There is a physical limit of how thin an injection molded wall can be manufactured. Depending on the length of the preform this may lie between 1 and 1.5 mm. However, injecting in this thin gap requires raising the injection and holding pressure beyond the recommended limits and there are operational consequences that make production more difficult.

[b]Chiang, Conger, Mainquist, "Integrated Approach to Lightweighting a Blow Molded Consumer Package," Proceedings of Annual Blow Molding Conference of the SPE, October 2006.

Figure 11.11 The bottom layer of a pallet is subjected to static load over a period of time. Picture courtesy of Clorox Technical Center.

Two new base designs have been developed that reduce weight and also streamline the injection process to some degree. The so-called Cappello design adds a different hemispherical shape to the base of the preform, reducing overall base weight and allowing faster injection speeds. The Eco Design by Husky Injection Molding Systems achieves similar results by replacing the hemispherical shape with a conical design. Both approaches claim weight reductions of 2–4% of the preform and bottle weight. In the ever more competitive world of PET bottles these can add up to significant savings (Fig. 11.13).

Special Machines

The proliferation of the 500 mL/20 oz PET bottle has led machine manufacturers to build machines that are limited to this container size. These machines are built on the frames of machines for bigger bottles but have more cavities. Mandrel pitch is limited to 38/40 mm as the 27/28 mm neck finishes do not need a larger pitch and the tighter spacing makes oven usage more economical. Clamps have shorter opening strokes and allow smaller heights only. This reduces the weight and space requirements of the clamp assembly making the machines more compact.

Figure 11.12 Lightweight sparkling water bottle longitudinal design features add to the top load performance of the bottle and can be aesthetically pleasing as well. Picture courtesy of Nestle Waters.

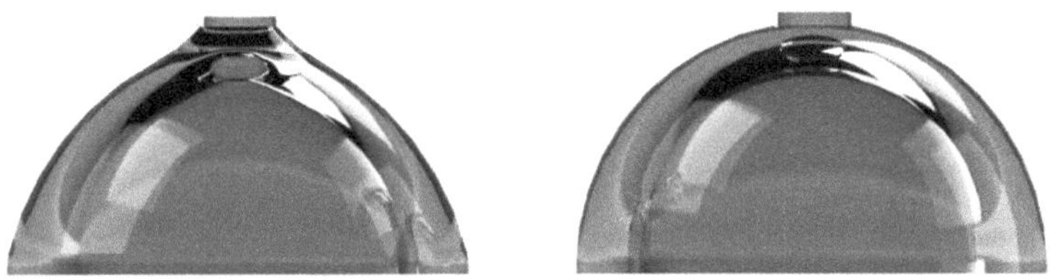

Figure 11.13 Cappello design reduces material on the outside of the preform end, adding some on the inside. As a result material savings of 2–4% can be accomplished. The design is compatible with both CSD and flat water bottles. Picture courtesy of Concordia Development S.r.L.

11.5 Resin Prices

Supply and demand control resin prices just like any other commodity in a market economy. PET has the added complication that polyester fiber with a global market volume many times that of PET uses the same raw materials - ethylene glycol and purified terephthalic acid (PTA). Supply of

PET is therefore often driven at least in part by the situation in the fiber market. While PET has been growing at double the growth rate of any other plastic, this growth has also attracted new capital investments including conversions of fiber plants to PET plants.

End users are most interested in low prices while resin producers are trying to maintain their margins. Converters are in the middle, balancing their resin costs with the pricing to their customers. To stabilize fluctuations, many companies have "banding" agreements with their suppliers and customers: as long as PET pricing, as published by various magazines and specialty services, stays within a defined range, they will pay and charge a preagreed amount. Only when this price exceeds the band will pricing be changed for both sides.

When buying resin, processors are well advised not only to look for competitive pricing but also to consider factors such as stability of supply, constant material quality, and customer service. Larger plants routinely check incoming resin for IV and AA content to ensure a smooth-running operation.

The resin market has changed significantly since this book was first written. Then Eastman Chemicals was the largest producer but they have since totally abandoned the market. Meanwhile, producers outside the United States have bought and started production facilities within the United States and many processors use a combination of domestic and foreign resin. While PET is not as dependent on raw material pricing as other materials, it does consume large quantities of energy during production and solid- stating. As a result, PET is less dependent on price movements in the underlying resource materials, namely crude oil. This explains why PET did not increase in price as much as the polyolefins: high-density polyethylene (HDPE), polypropylene (PP), low-density polyethylene (LDPE) when crude oil prices skyrocketed a few years ago. Prices in 2012 were close to par with PET higher. But in Jul. 2013 a divergence occurred, while polyethylene (PE) prices kept on climbing, PET started to go lower. This could have been caused by fluctuations in the fiber market or supply issues with PE. In any case, when oil prices started to slide in mid-2014 PE was already $300 higher than PET and the gap has narrowed slightly to $200, which was to be expected as PE could take better advantage of lower resource costs (Fig. 11.14).

If this trend continues, it bodes well for PET. It is easier to scale up, preferred by consumers because of its clarity, and offers weight savings in many applications.

The high energy input into its processing gives production facilities with low energy costs, such as parts of the Middle East, significant advantages and we can expect more dramatic changes to the make-up of PET supply over the coming years.

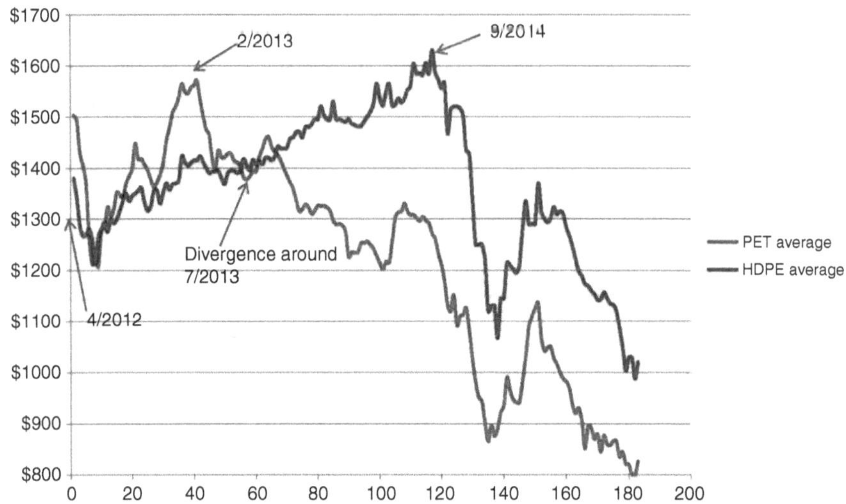

Figure 11.14 HDPE and PET prices have diverged over time to give PET a price advantage. While the shown prices may not reflect processor pricing in any geographical area because of difference in shipping costs and taxes, the graph clearly shows the trend.

11.6 Bottle Production Costs

There are two components to the overall bottle cost:

- Cost of preform, depending whether purchased or self-produced.
- Conversion cost from preform to blown bottle.

Preform Costs

Preform cost depends mainly on the following:

- gram weight
- preform wall thickness
- volume of orders
- neck size

Preform producers often quote the resin price separately adjusting it to fluctuations as the need arises. The resin proportion of the preform cost may constitute from 50% to 85% of total costs with heavier preforms making the top range (Fig. 11.15).

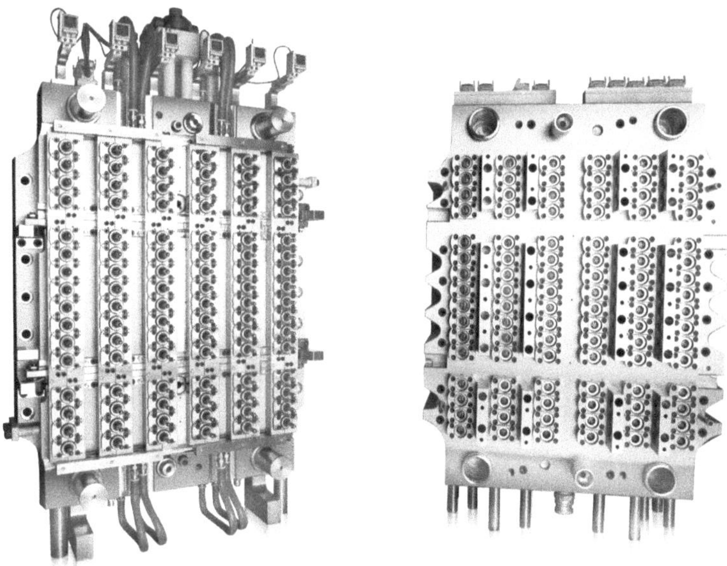

Figure 11.15 High-cavitation tools like this 96-cavity tool are used to produce most of today's preforms. Picture courtesy of Mold Masters.

Economy of scale plays another significant role in determining costs. Up to 216-cavity preform tools are currently in use providing a significant advantage over smaller tools if they can be utilized for a large part of the production year. High-capacity injection systems represent a significant capital investment but when spread over the hundreds of millions of preforms, which they produce annually (exact output depends on wall thickness—see next point) the cost penalty per preform is very small.

Modern injection machines for preforms come equipped with three or four postmold cooling stations that move into the molding area via servo motors. As soon as the skin of the preforms is cool enough to be ejected, they enter water-cooled take-out tubes (either directly or through an intermediate plate) that bring their temperature down to a value that minimizes blemishes. Some machines also blow air through custom tubes into the inside of the preforms thereby cooling the warmer gate area. Compared to running without postmold cooling this device cuts cycle time by about 1/3.

Output is not so much dependent on total weight of the preform but rather on the wall thickness. For example, comparing a 16 g to a 48 g preform with the same wall thickness, only the injection time would vary by approximately 2.5 s. However, a change of wall thickness from 2.2 to 4 mm would incur a cycle time penalty of up to 11 s. Hold and cooling time constitute a larger part of the overall cycle time than injection time independent of the system and machinery used. This is because cooling and hold time are proportional to the square of the preform wall thickness.

Neck size determines the maximum mold cavitation that will fit in a certain machine frame. Great strides have been made to locate the tooling components of individual preforms closer together. For many years, a 300-ton injection machine was needed to run a 48-cavity mold. Today, the same machine can accept a 72-cavity tool, boosting output per capital cost and floor space. However, a 45-mm neck can only be run in 48 cavities on this machine and only if it is set in the same tight pitch. This results in less output per machine tonnage thus increasing cost of manufacturing for wide-mouth containers.

Bottle Costs

The components making up total bottle production costs (excluding preform costs) are summarized in Fig. 11.16.

The relatively large proportion represented by machine investment is what saddles converters with high interest charges. The overhead figure depends very much on the company structure and is shown mainly to include a detail which is often missed in discussions about bottle pricing.

Energy costs deserve some special attention. Not only since the Kyoto Accord has public attention focused on this area of industrial production, but also careful planning and choice of equipment offer savings in this area.

The typical energy costs associated with blow molding (Fig. 11.17).

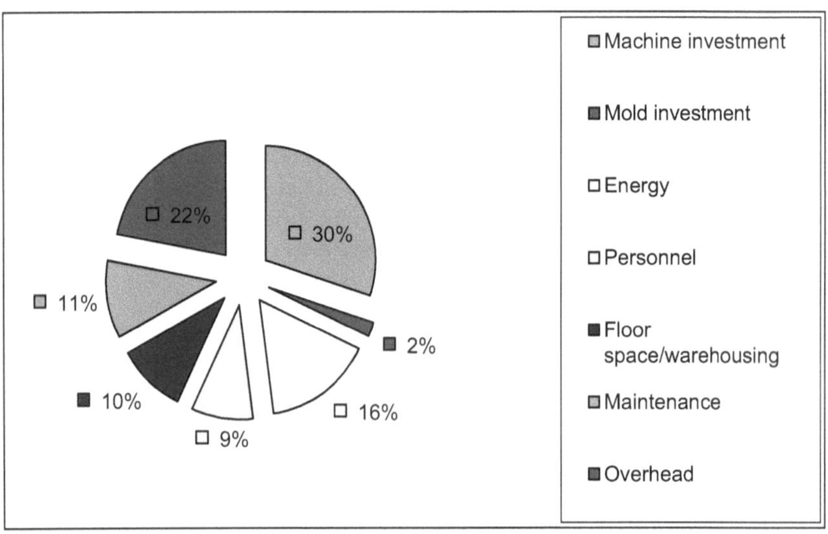

Figure 11.16 Typical bottle production costs.

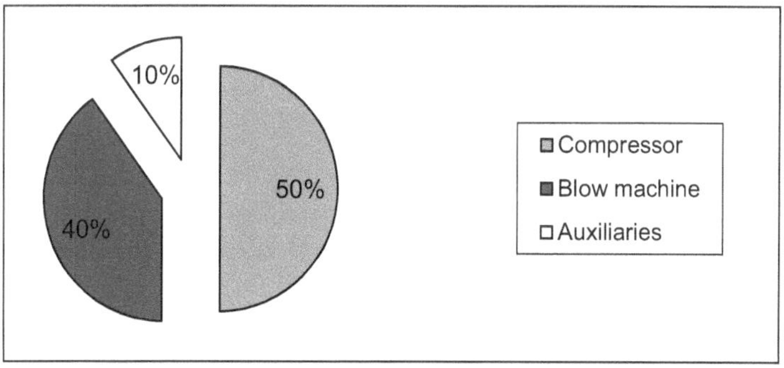

Figure 11.17 Typical energy costs in bottle production.

As can be seen, energy for high-pressure air can contribute as much as 8% (50% of 16%) to total bottle cost. The pressure setting plays a significant role in determining compressor power consumption. If a bottle can be successfully blown at 35 bar it should not be blown at 40 bar and companies are well advised to test each and every bottle for minimum pressure requirements and adding this parameter to the set-up sheet of each bottle. As described in Chapter 4 dead air volume has to be seen in this context and should be carefully investigated before purchasing new equipment.

11.7 Starting with a New Product

There are many aspects worth considering before embarking on PET bottle production. More than anything, the end product has to be cost-effective and analysis often suggests that buying bottles is cheaper than making them in-house. There are, however, other considerations that might outweigh this. Stability of supply is of major concern and being dependent on a bottle supplier means giving up control.

Bottle Demands

Proper analysis starts with plotting expected or actual demand over a calendar year. Many products have seasonal demand peaks and troughs as shown in Fig. 11.18.

There are different solutions to satisfy fluctuations in demand:

- Buy preforms and blow bottles only:
  - Pro: Low capital investment, lowest demand on process technology.
  - Con: Preform supply may limit design freedom, supply not guaranteed.

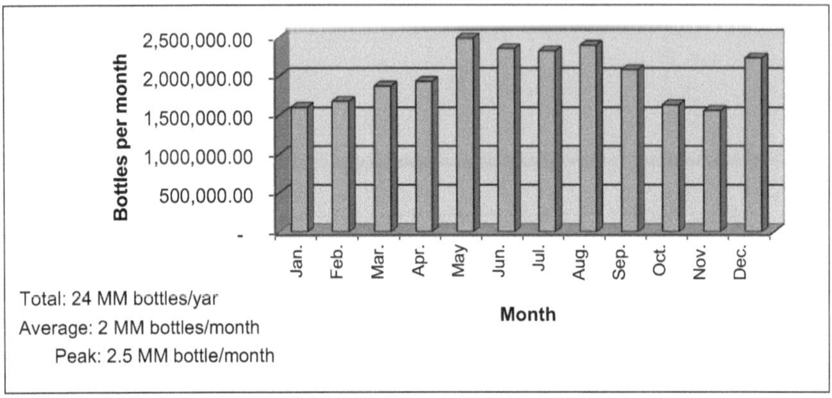

Figure 11.18 Seasonal bottle demand.

- Build machine output for lowest demand and buy bottles as needed.
  - Pro: Lowest capital investment; machines running continuously.
  - Con: Availability of bottles not guaranteed; bottle design limited to standard bottle, bottle price might be higher in peak demand months.
- Build machine output for highest demand and 7 days/week production, switch to 6 or 5 days/week production when demand drops off.
  - Pro: Custom bottle possible; control of supply.
  - Con: Highest capital investment; machines may be idle for part of the year.
- Build preform machine output for average demand and produce preforms for peak demand in low demand months or buy preforms in high-demand months.
  - Pro: Medium capital cost at full control of supply.
  - Con: Must choose two-stage process; blow machine may be idle for part of the year; preforms may not be available when needed.

Process Selection

There are several criteria that should be evaluated:

- Output: Here are some general guidelines:
  - Up to 20 MM/year: All processes
  - 20–120 MM/year: Two stage and integrated two stage
  - Over 120 MM/year: Two stage

- Demands on the bottle:
  - Highest demand: Two stage and integrated two stage
  - Lower demand: All processes
  - Rectangular bottle shape: Some bottles easier in single stage
  - Hot-fill: All processes but less experience in single stage
  - AA content: All processes but two stage more proven
- Aesthetics:
  - Must be without scratches: Single and integrated two stage
  - Thread orientation: Single stage automatically and two stage with special device
  - No neck support ring: All processes, but much easier with single stage

Machine Output Considerations

Here are some important "cause and effect" relationships affecting machine output:

- Bottle design dictates preform design. The more requirements the bottle has to fulfill, the more important correct preform design becomes.
- Preform wall thickness dictates cycle time, to a lesser extent in RSBM.
- Thickness of bottle base often dictates necessary cooling time and with that, cycle time and output.
- Neck finish diameter dictates number of cavities in injection mold tool. Also limits process choice: many two- and single-stage machines are limited to a maximum neck size.
- Bottle diameter or width determine number of cavities in single stage.

To compute total machine up-time it is important to consider the number of stock keeping units (SKU) planned. One bottle type may have several SKUs for different label configurations, products filled, or closures. Each SKU usually requires an interruption of the manufacturing process at some point and thus a fall in machine utilization. Changeover times vary greatly between machines of different processes and even different manufacturers. And the production machines themselves need regular maintenance in addition to unexpected stoppages that have to be factored in.

Most plastics plants operate 24 h/day, 250–365 days/year. All processes where PET is injection molded should run for 24 h/day because it can take up to 3 h before the process has completely settled and daily start-up costs would be prohibitive. RSBM machines on the other hand, are less sensitive and could run for one or two shifts only. However, high capital investment, especially for rotary machines, usually dictates 24 h/day production.

Decisions will be volume-driven in most cases. The smallest RSBM machine that is still economical would be two cavities and will produce between 2000 and 3600 bottles/h. This translates into 15–27 million bottles annually or a lesser number if utilization is less than 100% (which it is in most cases). If the machine needs to be changed over twice a week and taking into consideration seasonal demand fluctuations, utilization can easily fall to 60–70% reducing output to as low as 9–16 million bottles/year. This should be the number of bottles to look for when a machine purchase is being considered. It will be difficult to produce smaller volumes economically unless special conditions exist.

Volumes in the hundreds of millions of bottles are much easier to contemplate and many fillers have adopted blow molding as the high transportation costs of blown bottles have dug deep in their margins.

Planning for single stage is different as the output per cavity is much smaller. A typical machine will make 250 bottles/cavity per h (see Chapter 9, Section 9.9). A six-cavity machine could nominally make 12 million bottles at 100% and 8.4 million at 70% utilization. The higher tooling costs compared to stretch blow only must be counted against the purchase of preforms.

Transportation costs are a much smaller issue with preforms and many fillers leave this more complicated technology to converters. Because of the high output of even small injection machines, minimum requirements for economical production are usually higher except for wide-mouth applications. It is difficult for small producers to compete with converters running high-cavitation machines but on the other side of this argument are cases where the capital cost of a custom preform tool in high cavitation is too large to be justified and a smaller tool at the filler may be run successfully.

Process Flow

Converter plants generally use three types of packing to prepare bottles for shipment: palletization, shrink-wrapping, or cartons. Value-added services may include printing and/or labeling. In filling plants with bottle production there are several ways in which preform production, bottle production, filling, capping, labeling, and packing can be combined. Each comes with its own set of capital costs, plant layouts, and personnel requirements. The challenge is to find the right compromise between packaging diversity and operating efficiency.

Whether converter or filling plant, some type of buffer between the blow molding machine and downstream equipment is needed to avoid frequent machine stoppages. These lines must be designed so that each piece of downstream equipment has a higher output rating than the machine immediately upstream of it. Only then can temporary accumulations possibly be worked through the line. It is also paramount during the design process to consider the most likely stoppage frequencies and durations of downstream equipment. Based on these assumptions, sufficient buffer solutions can be implemented.

The simplest form of a buffer is a long conveyor belt transporting bottles at a higher speed than the upstream equipment delivers them. In times of normal operation, the conveyor will be only partially filled. Container densities on conveyor belts may range from 50% to 70% leaving the remainder for accumulating capacity. When the downstream machine stops, the conveyor allows the upstream machine to continue to operate by filling up with bottles. Buffer time depends on throughputs, length of conveyor, and bottle dimensions (Fig. 11.19).

Conveyors may make U-turns or transport bottles above machinery to save floor space. It is not unusual to have conveyors that are hundreds of meters long but the capital investment may be prohibitive for these kinds of solutions.

When longer buffer times are required, other solutions come into play. For any such system, sensors detect when the downstream equipment causes the conveyor to reach capacity and the upstream equipment must be stopped or bottles diverted.

Bottles may be diverted into storage silos or palletizing/depalletizing systems. Silos must be designed to avoid bottles being crushed and often have a series of baffles breaking the total distance a bottle falls into small steps. These are the most cost-effective solutions but may not be suitable for all bottles. Some bottles may need air cooling before they fall into the silo. Silos work in combination with unscramblers that deliver bottles back onto the conveyor when the downstream equipment is working again. Bottles from the upstream machine and the silo are merged and may exhibit slightly different process characteristics (Fig. 11.20).

In another approach, all bottles go directly to one or more silos and are always fed from these to the next piece of equipment. This reduces the need for long and costly conveyors and delivers all bottles with less variance in process history. Conveyors feeding silos can be less costly bulk

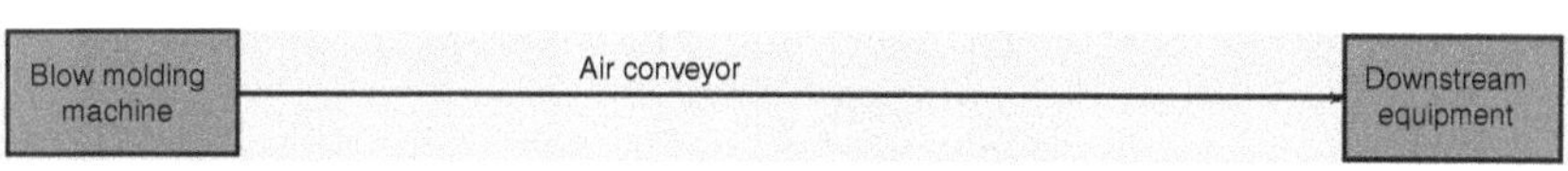

Figure 11.19 The 29 mm closure offers more consumer convenience at almost the same weight as the 26.7 mm closure. Courtesy of Corvaglia Mould AG.

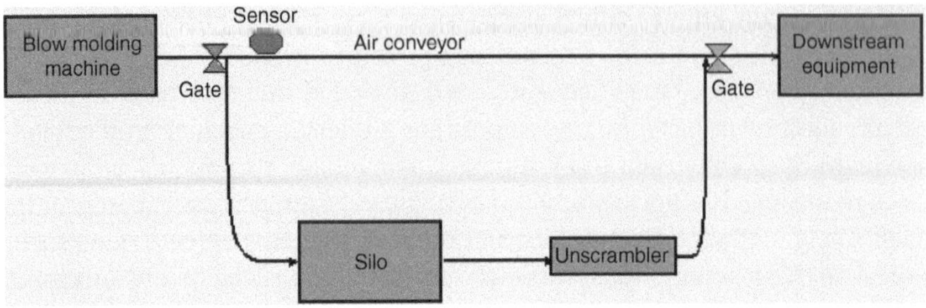

Figure 11.20 Here a silo is used to take overflow bottles and deliver them later to the downstream equipment.

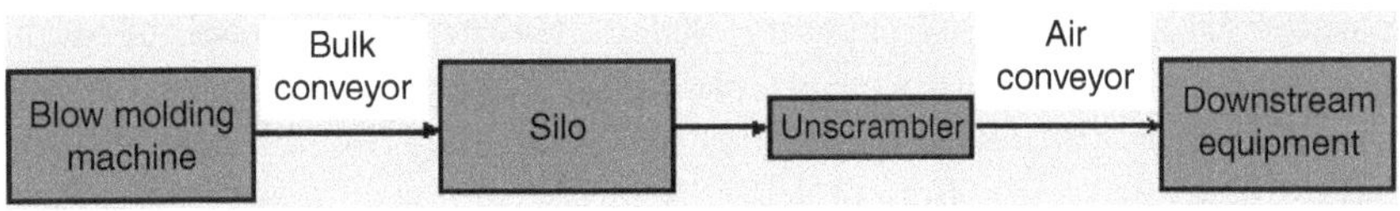

Figure 11.21 Here all bottles go into the silo reducing cost and different heat history of bottles.

conveyors. The disadvantage of this solution is that from an operational point of view it is counter-efficient to scramble the single-file bottle stream coming out of the blow machine. Every time bottles are brought back to single file bottle damage and jams are a distinct possibility (Fig. 11.21).

Accumulators of various designs can increase the 3–10 min of buffer time provided by a long conveyor up to the 30–45 min required for a product changeover at the filling line or similar.

Palletizers/depalletizers are a popular type of accumulating system and have the advantage that bottles stay oriented and are not subject to damage. These systems can be designed to handle any kind of bottle and buffer time (Fig. 11.22).

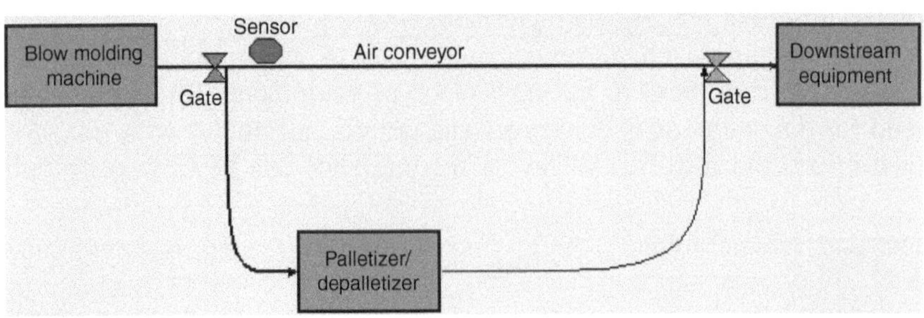

Figure 11.22 Palletizers keep the orientation of the bottles and are less likely to cause any damage.

Another approach to process flow is to fully synchronize various functions in a so-called "Monobloc" configuration. A good example is a form-fill-seal system now offered by a number of suppliers. Here the blow machine is linked to both filler and capper and all three machines work at the same speed. A rinser may be installed between blow machine and filler to cool down petaloid bases of CSD bottles that otherwise might not be able to take the internal pressure during and after filling. A great advantage of these systems is that bottles are always held and spaced in a controlled manner avoiding any handling problems associated with feeding of bottles into machinery as well as reducing labor costs. Some of these systems also offer dust removal from preforms with ionized air and/or UV treatment of the necks besides vision inspection system for faulty preforms. Labelers may be added to the block to complete the bottles for packaging. These systems are well-suited for the aseptic filling process where a minimum amount of operator intervention is paramount. The downside to this approach is of course that the whole line goes down when one component stops. Hence meticulous attention to maintenance schedules is required and the rejection of faulty preforms. Monobloc systems carry a premium over distributed systems that must be weighed against the savings in labor costs (Fig. 11.23).

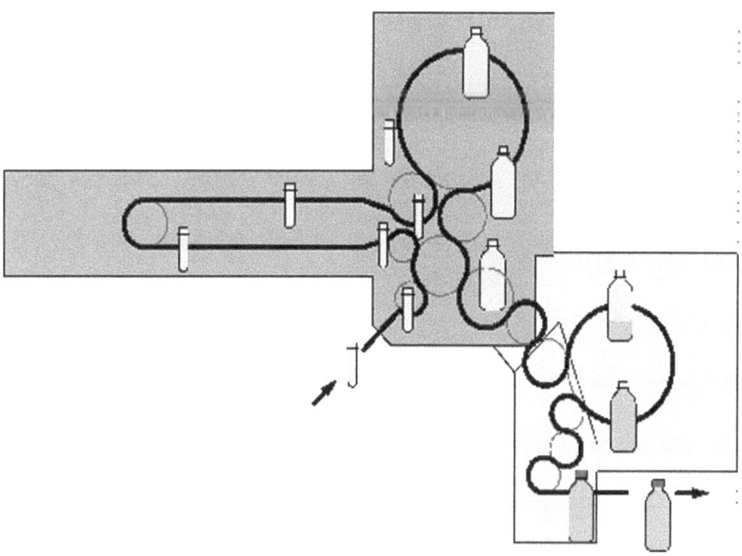

Figure 11.23 Blow-Fill-Seal systems save space, handling, and labor. Diagram courtesy of KHS Corpoplast.

Putting it all Together

Careful planning is essential for success in today's highly competitive marketplace. The decision to take production of bottles and/or preforms in-house should be based on realistic data rather than mere assumptions. Fillers need to consider that bringing these technologies onto their production floor requires a long-term commitment to constant innovation to stay abreast of ever-improving technology. A checklist to use as a guideline for the planning process is listed here:

- Calculate overall and peak demand based on available data
- Add projections for 2-year growth
- Decide on custom or standard bottle
- Check availability of preforms for custom bottle
- Obtain prices of preforms and bottles from various converters
- Negotiate with converters and secure supply during peak demand times
- Obtain details of transport costs to factory
- Decide on inventory levels
- Determine cost per purchased bottle
- Obtain preliminary quotes from machinery suppliers
- Assume operating efficiencies for:
  - Two-stage blow molding: 97%
  - Injection molding: 98%
  - Integrated two-stage process: 96%
  - Single-stage process: 95%
- Calculate number of mold changes per month
- For the first year, double the changeover times listed by manufacturers
- Calculate downtime by taking into account changeover times, regular maintenance schedule, and unexpected machine stoppages
- Add 1–2% to downtime for every SKU
- Calculate effective machine utilization
- Take learning curve into account
- Have manufacturers requote equipment, if necessary
- Obtain quotes for all auxiliary equipment (for calculations refer to Chapter 13)
- Decide on storage solutions for preforms and bottles
- Obtain quotes for various solutions
- Calculate total capital outlay
- Calculate manpower required and customary wages

- Check resin costs and payment terms
- Calculate cost per bottle for each manufacturing model
- Calculate cash flow requirements based on payment terms to suppliers and customers as well as inventory levels
- Decide on best price/performance ratio model

It will take 2–4 months to work through this process and management at all levels must be involved in discussing the profound effects self-manufacture of preforms and/or bottles will have on the company structure.

11.8 Recommended Laboratory Equipment for RSBM Plants

Strict quality control of both preforms and bottles is paramount for an efficient operation. Skids of produced bottles must be tagged with production dates and time in order that quality personnel can easily locate and "quarantine" them when defective bottles are found. Efficient communication lines between quality control and manufacturing keep plants running smoothly and allow problems to be solved in a timely manner.

Capital and operating costs of a quality control department should not be underestimated when planning production. Many end users demand zero defect procedures effectively downloading their inspection requirements to bottle producers. QC personnel must be properly trained and clear procedures written up. Quality requirements are too varied across the spectrum of PET bottles produced to give more detailed guidelines but industry associations such as those for bottled water or CSD are an excellent resource to use.

Basic and optional equipment for quality control functions is given in Table 11.2. The amount of tests required depends on the bottles produced as well as stipulations by end users. Many of the devices listed allow direct connection to a computer for data input. QC personnel take a reading and then press a foot switch to transfer the data to preset spreadsheets. One step up from this approach are measuring machines that take bottles from crates and perform a series of measurements automatically, delivering written reports. These can be attached to shipments to customers (Fig. 11.24).

11.9 Western Versus Asian Machinery

It has not been that long ago when the machinery trade between East and West went only one way, and that was to the East. Asian machinery (in this context always with the exception of Japan), although cheap, lacked

Table 11.2 Basic and Optional Equipment for Quality Control Functions

Equipment	Function
Caliper	Preforms and bottles
Height gauge	Preforms and bottles
Perpendicularity gauge assembly	Preforms
Finish gauge	CSD preforms and bottles
Depth gauge	Fill point for bottles
Tape measure	Bottle diameter
Polarized light viewer	Preform examination
Syringes and accessories	Fill bottles with liquid to an exact level
Digital scale	Weight and volume
Top load tester	Top load for bottles
Thickness gauge	Preforms and bottles
Hot wire cutter	Bottle section weights
Concrete basin with drain	Drop test for bottles
Environmental chamber	Controlled aging of bottles
Thermometer	Temperature check for tests
Hypodermic needle	Carbonation level check for CSD
Pressure gauge/transducer	Carbonation level check for CSD
Gas chromatograph	AA content for water and CSD
Gel permeation chromatograph	IV determination
Burst tester	CSD

in many respects the technology and sophistication that are required to be competitive in the Western World. Taiwan was the first Asian country to break out of the mold and deliver machines that could rival to some extent, Western standards. Other countries, buoyed by domestic demand ramped up their export business and today machines for plastics production come from virtually all countries of the region. That is not to say that they are dominating any sector except one: two Japanese companies produce upward of 85% of one-stage machines. In other sectors Asian machines mostly play a much smaller role finding markets for small, unsophisticated machines that compete only on price.

For the Western converter that is considering Asian blow molding machinery because of the sizable price discounts, this chapter is meant to provide a checklist of some sort that outlines all considerations a prospective buyer should keep in mind. In the end, a low price helps with a current shortage of capital but is no guarantee of long-term low operational cost savings. To better understand the decision drivers, it is well worth to find out why many Asian machines are so much cheaper than Western ones. Here are the common categories:

Figure 11.24 Fully automatic test machine performs a number of measurements and delivers reports that may be sent with bottle shipments. Photo courtesy of AGR International.

- Cheap labor: While this point is frequently made, labor in the overall cost of a Western machine is less than 15%. Significant savings are there but not enough to explain the vast price difference.
- Engineering in manufacturing and R&D is another area where Asian companies spend less as most machines are copies of proven, Western models.
- Infrastructure: Asian factories are often unheated, rudimentary affairs that are much cheaper to build and maintain.
- Materials: Asian companies often (although not always) buy outdated software, use lower-quality steel, counterfeit electronic components, and cheaper machine tools to manufacture machines.

- Quality of workmanship: Asian workers are often less qualified than their Western counterparts as Asian companies spend very little of upgrading their skills.
- Many machines lack costly safety standards, an important point that buyers should be especially aware of.
- Government incentives and tax breaks help especially Chinese companies to lower their cost.

From this very superficial comparison it is now possible to point out what potential buyers should be aware of. First and foremost, it is important to take steps to ensure that the machine or tool is specified as exactly as possible. Every part should be described in as much detail as is appropriate. During a visit to the Asian factory personal, connections should be made so that future communication is working well. Language barriers are still a major problem especially in China while Taiwan has reached standards similar to some of the European countries.

When the machine/tool is finished, a detailed acceptance procedure should be undertaken that was of course agreed upon beforehand. Buyers have very good leverage at this point in the acquisition process as the manufacturer has spent most of the money to build the machine but will get a large percentage of the total only after the machine has been shipped. During this time, manufacturers are therefore more prone to accept correcting problems that surfaced during acceptance.

A "CE" sticker is cheap to print and buyers should demand to see the CE-Certificate and get a copy. If there is an injury on a machine, insurance companies are all too eager to find the safety measures on a machine in violation of current laws, which will enable them not to pay!

Once the machine is in the customer's plant, issues of service and spare parts availability come to the forefront. Even the cheapest machine is not worth anything if it cannot run because nobody can start it or parts that have to come from Asia are not there. It is obvious that Asian companies lack in this regard but customers can help themselves if they make detailed lists of parts and find out where to source them. A previously agreed-on spare parts package can also make a difference.

Most Asian machines also run generally slower than Western ones. In stretch blow molding, the fastest Western machine now produces 2250 bottles/cavity per h while the fastest Asian machine used to be at 1800. That is 23% less or 3.5 million bottles per cavity in 7000 h production time. However, recently one Chinese manufacturer showed a rotary machine running at 2400 b/h, 150 b/h faster than their Western counterpart. That is the first that this has happened and demonstrates that the current imbalance between East and West may come to an end. In injection molding the difference is now 2 s (7.5–5.5 as the fastest cycle times) but that too may soon be history.

When considering costs per piece, the calculations show that actual capital cost occupies a large piece of the pie. This is unlike injection molding of preforms where the resin cost is 70–85% of the total, leaving only a small percentage for capital cost. In stretch blow molding, the preform cost goes in as a scrap percentage but capital cost may be as 30% of the total. Therefore, cheaper machines hold the promise of significantly reducing manufacturing costs. This promise can hold if the machines are able to produce bottles and if the bottles produced can be sold to a customer. Here are some graphs that establish the various relationships. They were derived from changing one parameter only while keeping all other inputs the same (Figs. 11.25–11.27).

Putting these numbers in context, one can get the following conclusions:

- Doubling of the capital outlay increases conversion cost by 40%.
- An increase of scrap rate from 0.5% to 5% increases conversion cost by 15% for the 12 g bottle and 34% for the 30 g bottle.
- A decrease in productivity from 95 to 80% increases conversion cost by 10%.

In summary, careful preparation including written and detailed specifications as well as good project management are important for any capital outlay, no matter what machine is being purchased. Only when all relevant factors are considered, can a rational decision be made. If a company does not have the resources to get this done in-house, outside help should be employed.

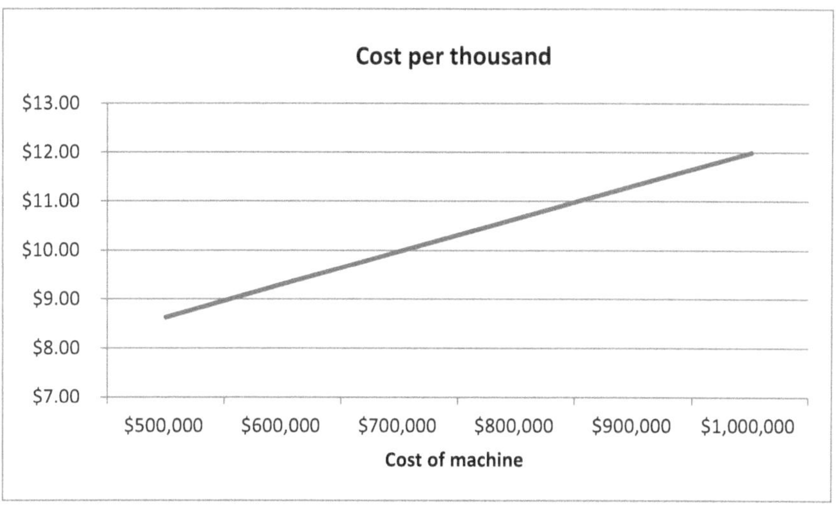

Figure 11.25 Capital cost has a major influence on part cost.

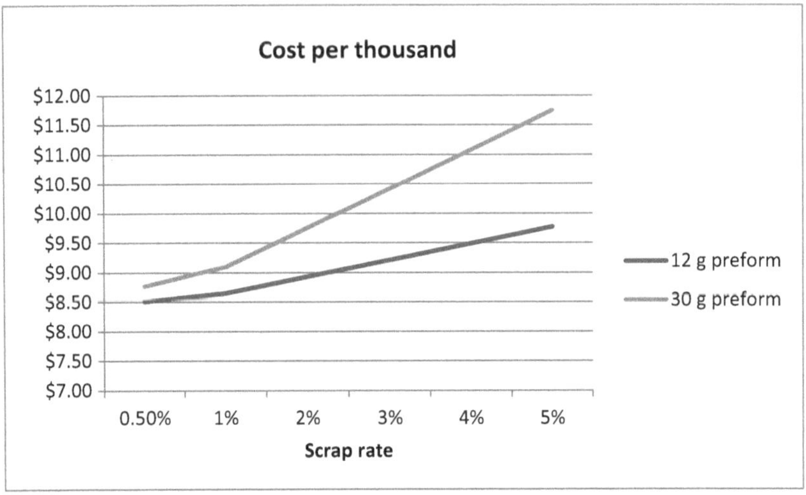

Figure 11.26 Scrap rate can also have a major impact on cost, especially as preforms get heavier.

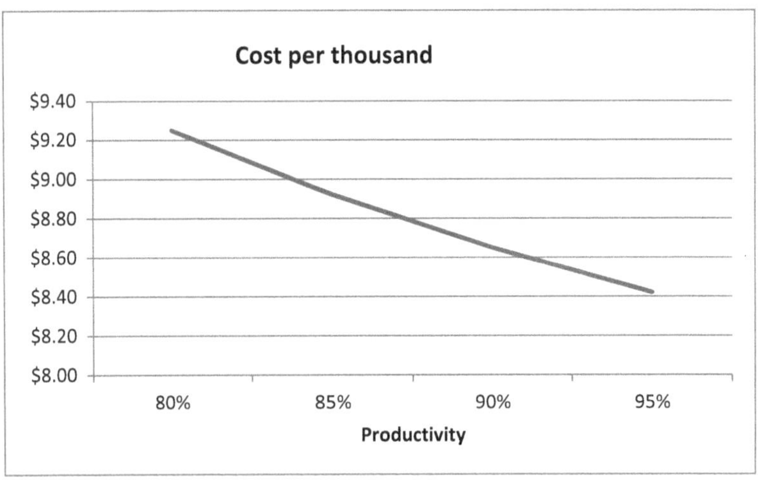

Figure 11.27 Productivity should also be considered when evaluating costs.

11.10 Bottle Storage and Transport

Considering all the effort and capital that needs to be invested in making high-quality bottles, it would be foolish to then squander these investments by not following guidelines to taking care of the bottles after blowing. PET is hygroscopic and will attract moisture during storage in high humidity environments that may lead to free volume relaxation and a change in bottle dimensions. Empty bottles may also shrink at elevated temperatures which can lead to capacity issues, especially fill levels. If they are exposed to direct sunlight, necks can get distorted (typically bent). To avoid these issues here are some guidelines that processors and fillers should follow:

- First-in first-out (FIFO) must be enforced at all times. This is harder than it sounds because forklift drivers have a tendency to go for the quickest route when they are under time constraints, which may not be the right one.
- South-facing warehouse walls must be painted white on the outside and empty bottles (especially CSD) should only be stored some distance away from those Use the empty space for other items. This is less of a problem with filled bottles; the liquid inside acts as a heat sink.
- Warehouse temperature should be monitored and fans employed to prevent heat build-up. Passive exhaust fans mounted on the roof are a good way to avoid the most excessive heat.
- Bottles should never be stored in tractor trailers in the summer time where temperatures can easily reach 60°C (140°F), very close to the glass transition temperature of PET.
- When loading trucks, bottles should be exposed to direct sunlight as little as possible.

PET is very resistant to acids but very weak against alkaline solutions that tend to accelerate stress cracking. Therefore, any exposure to these substances must be eliminated:

- Soaps containing NaOH cannot be used on any parts the bottles come in contact with.
- Lubricants must be free of stress-cracking agents.
- Bottles must be dry before they are packed into carton. If wet, not only can the carton fail but stress-cracking agents can egress from the carton causing bottle failures. This is accomplished by using bottle warmers that avoid condensation on the bottles after they have been filled with cold liquid.

Processors following these guidelines should have no problems with bottle failures caused by incorrect storage and transportation.

12 Preform Design for Single- and Two-Stage Processing

Chapter Outline

Preforms for these two processes are quite different and one made for one process will not work in the other. The reason for this becomes clear when we look at what is actually happening to the preform just prior to stretch blow molding (Fig. 12.1).

12.1 Two-Stage Process Injection Molding

In order to guarantee optimal flow of the molten polyethylene terephthalate (PET) into the injection cavities designers use ratios of gate wall thickness to body wall thickness. Ideally, one would want to keep the gate wall thickness as low as possible as this part of the preform does generally not participate to a large degree in the stretching process and stays as an often wasteful, thick disk in the center of the bottle. (In carbonated soft drinks (CSD) bottles, however, a thick center disk is required as a thinner, amorphous center would not be able to withstand the considerable pressure.) However, to make sure that PET can enter the cavity without too much shear stress, a ratio of 72–80% is used in most cases. That means that a preform with a 3-mm body wall thickness will have a gate wall thickness between 2.16 and 2.34 mm.

The exception to the rule is preforms with a body wall thickness of less than 2.5 mm. These have become quite popular with lightweight water bottles as a lower preform wall thickness reduces cooling requirements and speeds up cycle times. There is a minimum gate wall thickness of about 1.8–2 mm and gate to body wall thickness ratios of up to 100% may become necessary for these preforms.

Preforms are all at the same temperature when they are fed into the blow molding machine. In the heating section, they spin on mandrels while infrared heaters bring them to blowing temperature. This virtually guarantees

Stretch Blow Molding. http://dx.doi.org/10.1016/B978-0-323-46177-1.00012-3

Figure 12.1 Choosing or designing the right preform for a bottle is a crucial step for both functional and economical reasons. Photo courtesy of KHS Corpoplast.

that all preforms are at the same temperature circumferentially. One limitation of the process is that in order to protect the neck finish the first lamp in the oven section has to have a certain distance from the neck finish that is protected by a sheet metal shield. Therefore, the preform section underneath the neck support ring (NSR) cannot be heated to the degree that processors would want to. The material in this section forms the shoulder of the bottle where less material is needed and it is beneficial for overall weight savings to pull material out of this area. Preform designs for two stage take this into account by bringing the thinner neck section down to a point where material is then needed to form the wider body of the bottle.

When reheating preforms in a good blow molding machine, knowledgeable processors can actually bring the inside temperature of the preform to a higher level than the outside, achieving the optimal temperature profile through the wall. They do this by using high lamp settings and enough venting air in the oven to keep the oven temperature below blowing temperature. A longer equilibration time after heating is also helpful for this purpose. This is beneficial because the inner surface material of the preform has to stretch further than the outer one. By changing lamp heater outputs, oven fan cooling, and preform throughput operators can precisely adjust the temperature to an optimal level. For these reasons high blow up ratios up to 12 are possible in two stage resulting in very good properties such as carbonation retention for CSD bottles.

12.2 Single-Stage Process Injection Molding

In this process preforms and bottles are manufactured in the same machine. After injection molding, the preforms stay in the neck inserts and are shuttled to a conditioning station in some machines or directly blown in other machines. At this point in time the preform has only partially cooled down from the melting temperature of around 260°C (500°F). The cooling of the preform comes from the inside (injection core) and outside (injection cavity) during injection, resulting in the center of the preform wall staying warmest. This residual heat makes preforms from the single-stage process on average warmer than their two-stage counterparts.

A very beneficial feature of this process is the fact that the inner preform wall stays warmer than the outer one. This is due to the smaller core cooling area and lower core cooling water flow as explained in more detail in Chapter 10, Section 10.6. This feature allows the process a good degree of flexibility when it comes to oblong bottles.

Since the last part of the preform to receive hot material is toward the gate area so this part of the preform is always warmer than the area underneath the neck. Processors are therefore limited in how they can process these preforms. Designers compensate for this in two ways. First they use a lower gate to body wall thickness ratio, usually between 58% and 66%. This reduces wall thickness in the gate area allowing faster cooling. They also tend to have the greatest body wall thickness underneath the neck if possible. Quite contrary to intuition, a thicker wall in the PET preform in this process leads to a thinner wall in the corresponding bottle wall thickness. This is true because a thicker wall retains more heat and will then subsequently stretch so much more that a thinner bottle wall is achieved. But nonetheless, when the blow molding process is the slave of the injection process, process capabilities diminish.

12.3 Goals and Conditions

There are several goals every designer has at the outset of designing a preform. These are as follows:

- Creating the strongest possible bottle at the lowest weight, be it with respect to parameters such as top load or carbonation retention or to give consumers the feel of a well-made container.
- Even wall distribution in the sidewalls.
- Strong bottom corners that cannot easily be pushed in.
- Least amount of material in the base to still make a functional container.
- High injection and blow machine output.

Some of these goals may come in conflict with each other and often preform designs end up as a compromise between different requirements. In fact, designers are given certain requirements at the outset and they have to work within them. These are as follows:

- bottle design including neck finish;
- weight;
- application such as water, CSD, personal care, etc.; and
- process (single or two stage).

12.4 The Mechanics of Preform Design

Preform design has long been considered somewhat of a black art and even today most designers will consult their library of past designs first when they get a new project rather than going through some scientific evaluation. While it is true that preform designs often do not work out as anticipated there are a number of design considerations that will put any new design on a solid footing. Elements such as the state of the blow machine, the skill of the processor, and plant conditions all have an effect how a particular preform works out and are out of control of the designer. In any case, new designs need to be evaluated in unit-cavity tools before an expensive preform tool can be ordered.

All designs start with a neck finish as it will determine the overall shape of the preform. The inside and outside diameter of the neck are the starting point from where the designer has to choose from four basic shapes (Fig. 12.2).

The application will determine what stretch ratios to employ. As explained in Chapter 7, Section 7.2 the overall or planar stretch ratio (PSR) is the product of internal hoop and axial stretch ratio. Table 12.1 outlines typical ratios for different applications.

Very lightweight water bottles with small, 26-mm necks have forced developers to venture into ratios as high as 12.5 and these are employed today in some cases. The table values are more conservative figures that designers should stick to as safer starting points. It should also be noted that Table 12.1 reflects ratios employed most in the two-stage process. In single stage, where processors have less control over the preform heat profile, ratios seldom exceed 8, and are often well below that. That is not a problem in most cases as bottles produced in this process have thicker walls than their two-stage counterparts and the degree of orientation required is mostly not an issue. Therefore, the following discussion is relevant mostly for the two-stage process.

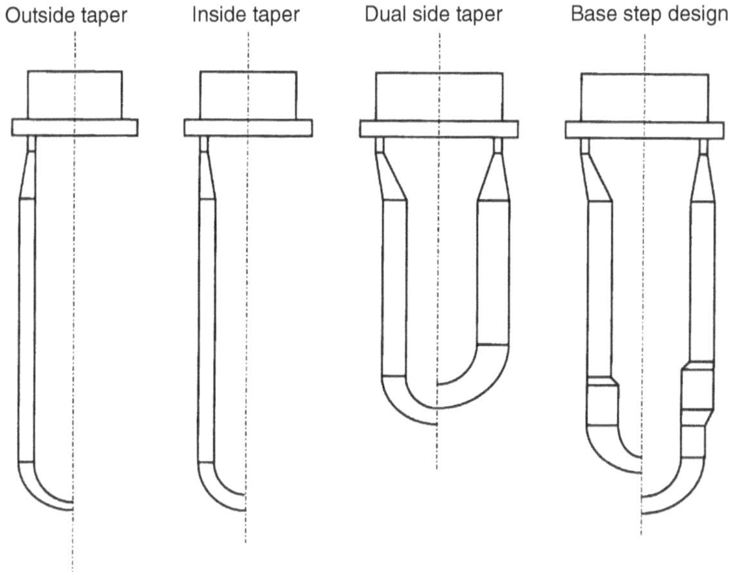

Figure 12.2 Preform shapes. These basic shapes are available for the designer.

Table 12.1 Typical Ratios for Different Applications

Planar Stretch Ratio	Application
11–12	CSD, oval, rectangular
10–11	CSD, flat drinks
9–10	Water, edible oil
8–9	Hot fill
8	Heavy bottles

The first rule to spell out here is that ratios in the axial and hoop direction should be at least a factor of 2, which means 4 for the hoop direction as this is a diameter comparison. This is to ensure a certain degree of orientation in both directions and is the reason that 8 is the lowest planar ratio listed. When higher ratios are required it is beneficial to expand both hoop and axial ratio by the same multiplier but this is not as critical.

After choosing a certain PSR the designer would start in determining the axial stretch ratio. Since the hoop stretch ratio is double that of the axial stretch ratio in the ideal scenario, the simple formula: PSR = $X \times 2X$ is solved as X = square root of (PSR/2). For example, a planar ratio of 10 would let the designer start with an axial ratio of square root of 5 or 2.3 and that would determine the length of the preform. He then would determine the best preform shape that will result in an internal stretch ratio of 4.6. All modern 3D computer-aided design programs let users calculate the weight of a created object and designers use this feature to determine the wall thickness of the preform once the basic shape has been established.

Bringing stretch ratios in-line with the application is not the only task of the designer. The next step is to determine where the thin neck parts should transform into the thicker body parts of the preform and how thick to make the preform base. For the first task it is important to correlate the bottle shape to preform wall thickness. All bottles have a shoulder sloped at different angles. In the shoulder region less material is needed to create a certain wall thickness and the transition from neck to body diameter in the bottle should be reflected in the preform. To better understand these relationships, here is an example.

The transition area from thin to thick on the preform is the distance from A to B. The transition area on the bottle from the neck to the body is the distance from A to C. The transition ratio for the preform is (A to B) divided by (A to D) while that of the bottle is (A to C) divided by (A to E). These two relationships should be very similar if not identical. In this example it is:

Preform: 14.11/89.16 = 15.8%
Bottle: 39/196.74 = 19.8%

If the transition ratio in the preform is smaller than in the bottle (as is the case in the example) a line can appear in the bottle where the thicker preform wall section ended up. If the difference is small (<5%) the process can usually be adjusted to avoid this issue.

If the transition ratio in the perform is larger than in the bottle, the bottle shoulder becomes weak and poor top load results (Fig. 12.3).

Also important in this regard is determining the distance underneath the bottom of the transfer ring that must be straight on the outside to allow grippers or other handling devices access. In most cases this distance will be 4–5 mm.

While preform walls are mostly kept straight both inside and outside must be slightly tapered to allow demolding. Designers use 20–40 degree-min for this purpose. In single stage, it can be beneficial to have a larger taper that allows more material in the area underneath the neck to facilitate a thinner bottle shoulder. Larger tapers are also employed in preforms for wide-mouth bottles. This is because if straight preforms were designed for necks of 63, 89, or 120 mm too much material would end up in the bottom of the bottle. Instead, these preforms are tapered to diameters that will transform into bottle bases with suitable wall thickness (Fig. 12.4).

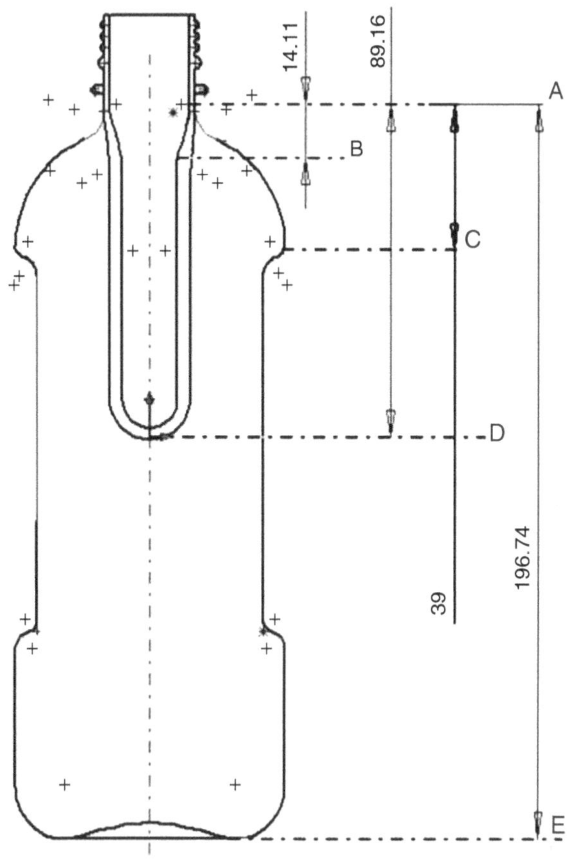

Figure 12.3 Bottle and preform.

12.5 Putting it all Together

Here is an example. I will design a preform for the bottle depicted in Fig. 12.5. The 28-mm plastic cap only neck finish weighs 5.1 g and will not be drawn. The weight will just be added at the end. Here are the design parameters:

Volume:	1 L
PSR:	10.5–11.5 for maximum oxygen protection
Weight:	30 g
Application:	Flat, flavored water

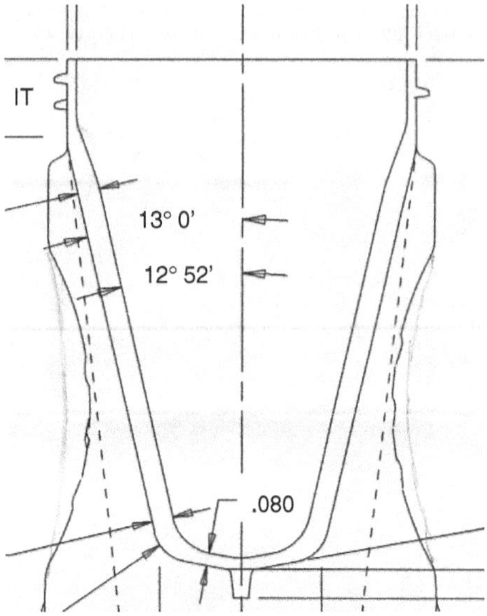

Figure 12.4 To avoid thick bottle bases wide-mouth preforms are designed with large tapers.

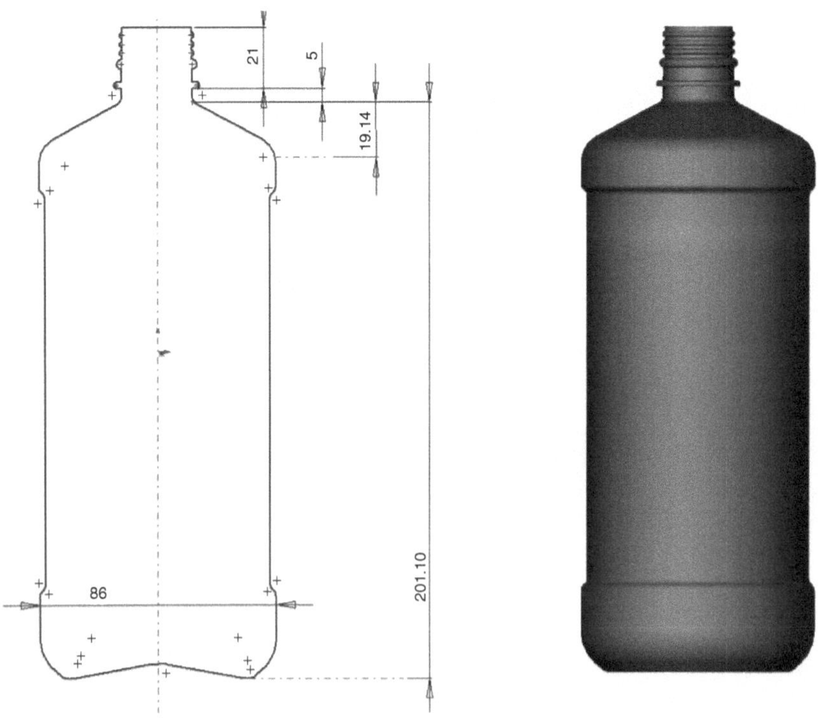

Figure 12.5 Water bottle.

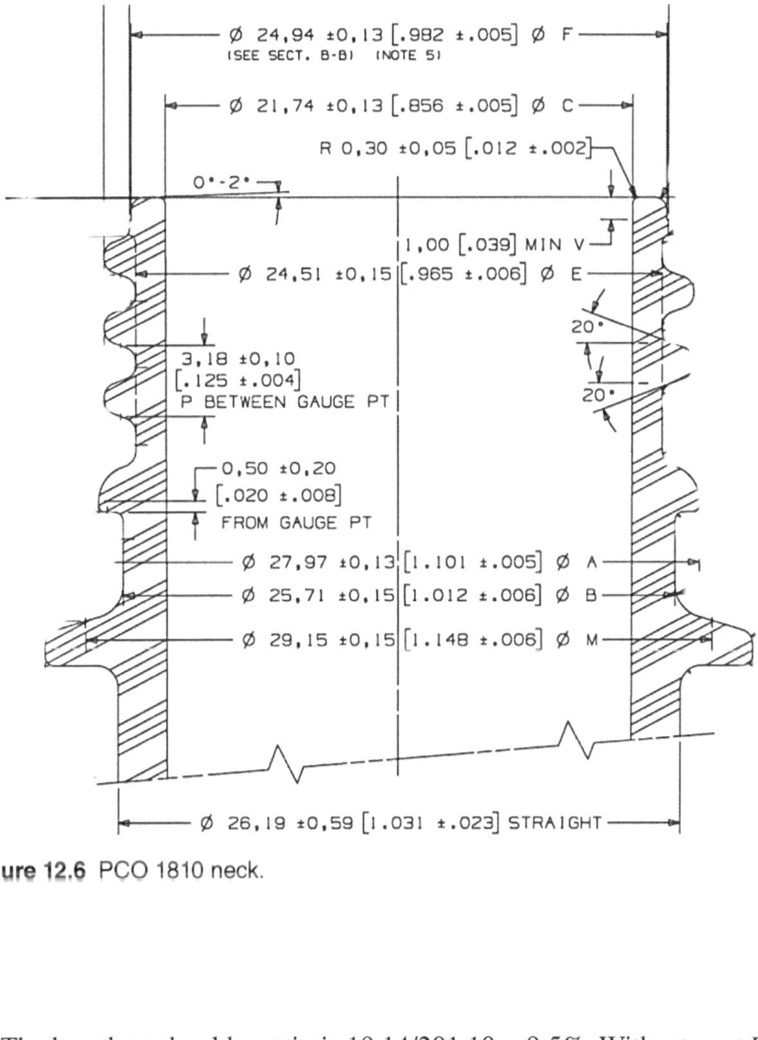

Figure 12.6 PCO 1810 neck.

The length to shoulder ratio is 19.14/201.10 = 9.5%. With a target PSR of 11, the axial stretch ratio will be the square root of 5.5 or 2.35 and the internal hoop stretch ratio 4.7. These are just target values that may change in the course of the design but they are a good starting point.

Preform design starts with the neck dimensions as specified in Fig. 12.6.

The outside dimension of the preform will be 26.19 mm as shown, the inside dimension 21.59 mm (the inside taper reduces it from 21.74 mm as shown). At this point, it is not obvious which of the design options will be used. Starting with a straight outside wall would be the first option to consider. Here are some of the initial calculations:

Axial stretch ratio:	201.10/2.35 mm = 85.57 mm
Transition length:	85.57 mm × 9.5% = 8.12 mm
Inside diameter of the preform:	86/4.7 mm = 18.3 mm
Resulting wall thickness:	(26.19/18.3 mm)/2 = 3.95 mm
Gate wall thickness:	3 mm
Gate to body wall thickness ratio:	3/3.6 mm = 83.3%

A wall of 3.95 mm will command sizable cycle time on the injection machine. Most designers try to stay within a 3.5–3.7 mm band for these types of bottles. There are now three choices if the target wall thickness should be 3.6 mm:

- increase the preform inner diameter (ID) resulting in a loss of hoop stretch ratio;
- increase the preform length resulting in a loss of axial stretch ratio; and
- reduce the preform outer diameter (OD) to 25.5 which would result in dual side taper design.

In this example, the third option will be chosen.

At this point it is not known what the preform weight will be and it can be fruitful to calculate it with the help of a spreadsheet (Fig. 12.7).

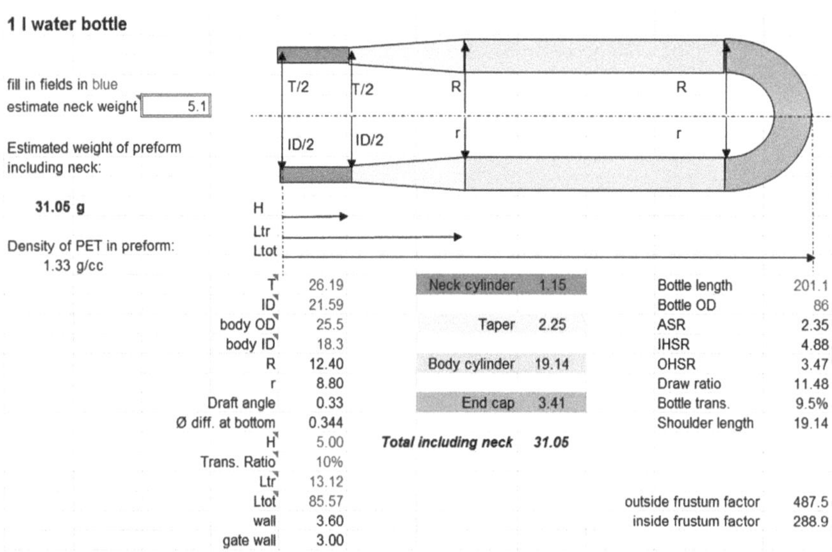

Figure 12.7 A spreadsheet may prove helpful to calculate weight before starting to draw the preform.

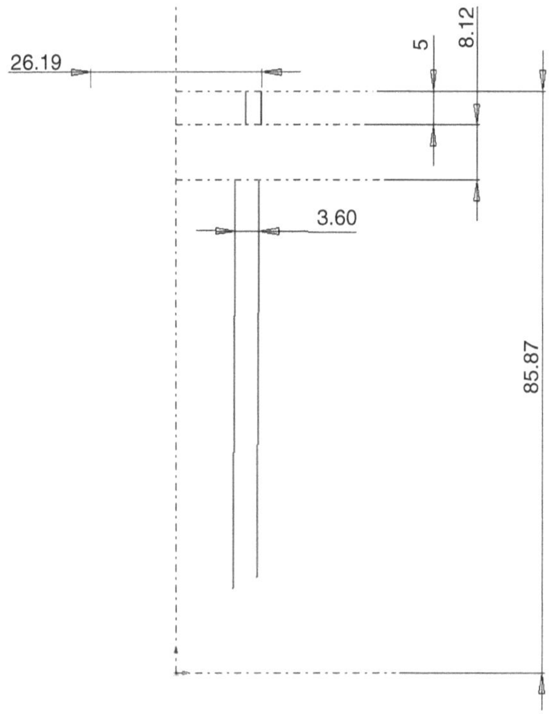

Figure 12.8 Preform for 1 L bottle, step 1. Most calculated values have been incorporated in this first step.

This calculation shows a weight very close to the target of 30 g, close enough indeed to start drawing. A first sketch may look like the one given in Fig. 12.8.

Notice that both body lines are in an angle of 20 degree-min. It is important to draw the tapers right away as they are difficult to change later.

It is now easy to connect the two lines coming from the neck with the two lines forming the body and create the transition zone. To create the bottom hemispheres and connect them to the body lines is a little more difficult because of the taper. Trigonometry reveals that the two body lines shift about 0.344 mm to the inside where they will meet the hemispherical shapes. This will result in radii of 12.4 and 8.8 mm. Since the gate wall thickness will be 3 mm it is now possible to draw the center lines of those radii. The 12.4 line starts at the bottom of the preform and the 8.8 line 3 mm above it (Fig. 12.9).

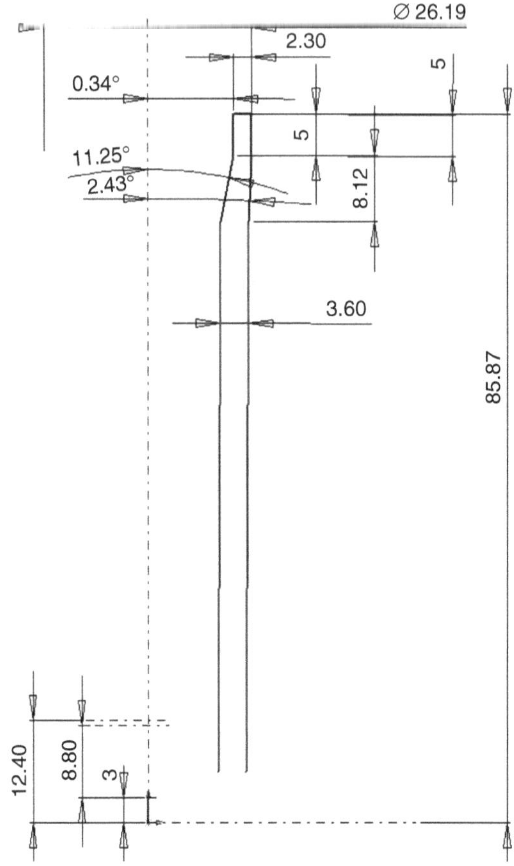

Figure 12.9 Preform for 1 L bottle, step 2. The transition area has been drawn as well as the center lines for the bottom radii.

When connecting the two radii with the bottom lines it is important that the lines perfectly match (Fig. 12.10).

The calculated weight is short of the neck weight of 5.1 g, which brings the total to 31.32 g, very close to the calculations in the spread sheet (Fig. 12.11). To bring the preform to the target weight of 30 g, here is a choice of

- reducing the preform length; result will be an increase in axial stretch ratio;

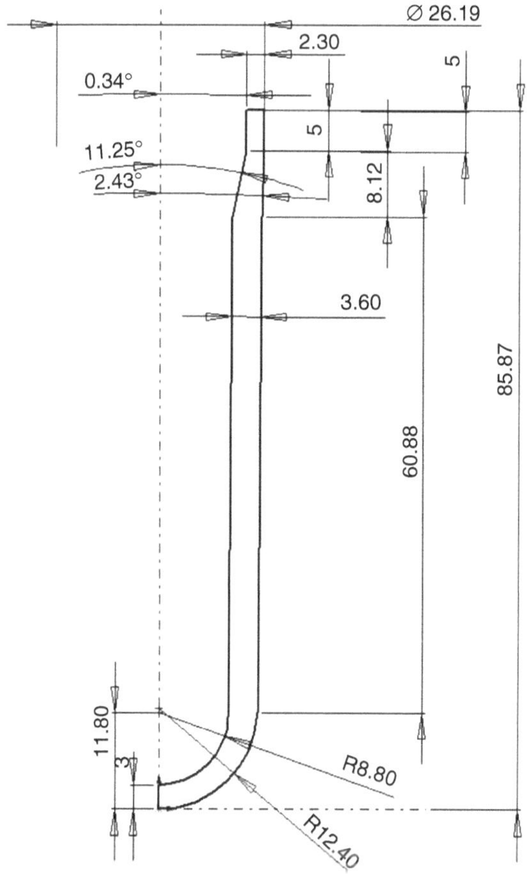

Figure 12.10 The completed preform.

- increasing the preform ID, result will be a decrease in hoop stretch ratio; and
- decreasing the external diameter.

Since the draw ratio is already at the upper design limit of 11.5 it seems best to increase the preform ID. Changing the numbers on the spreadsheet makes this an easy task (Fig. 12.12).

Fig. 12.13 shows the finished preform drawing with the changed ID and radii in the connection zones.

The final preform will be easier to injection mold and still have excellent barrier properties with a high draw ratio of 11.2. This design can now

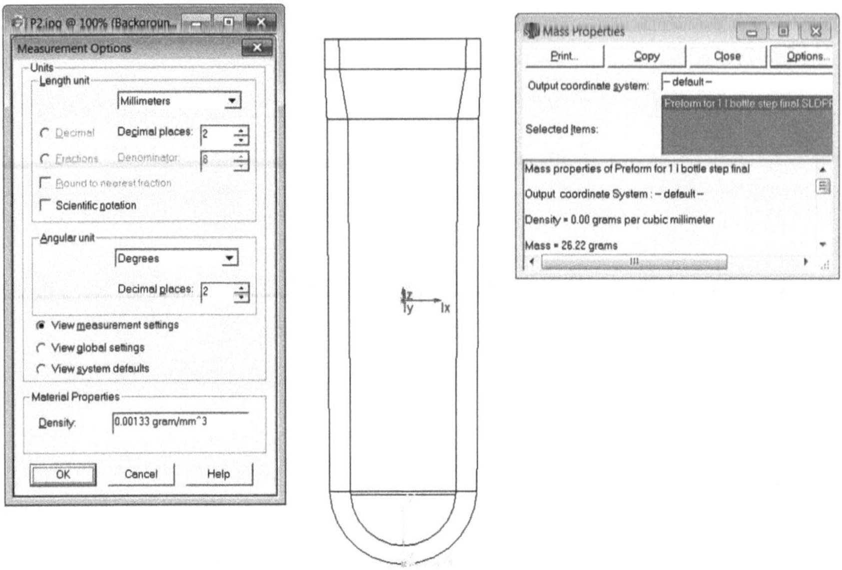

Figure 12.11 The software reveals a weight of 26.22 g at the PET density.

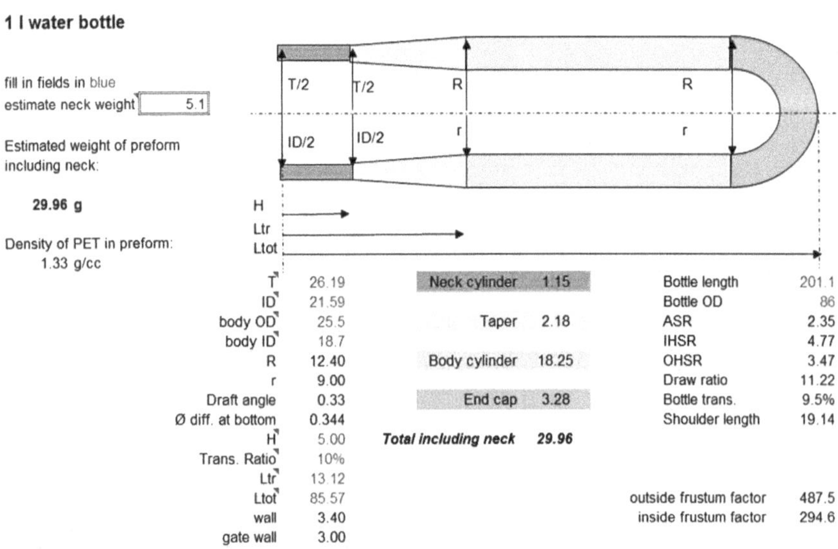

Figure 12.12 A small change in the body ID brings the preform to the correct weight.

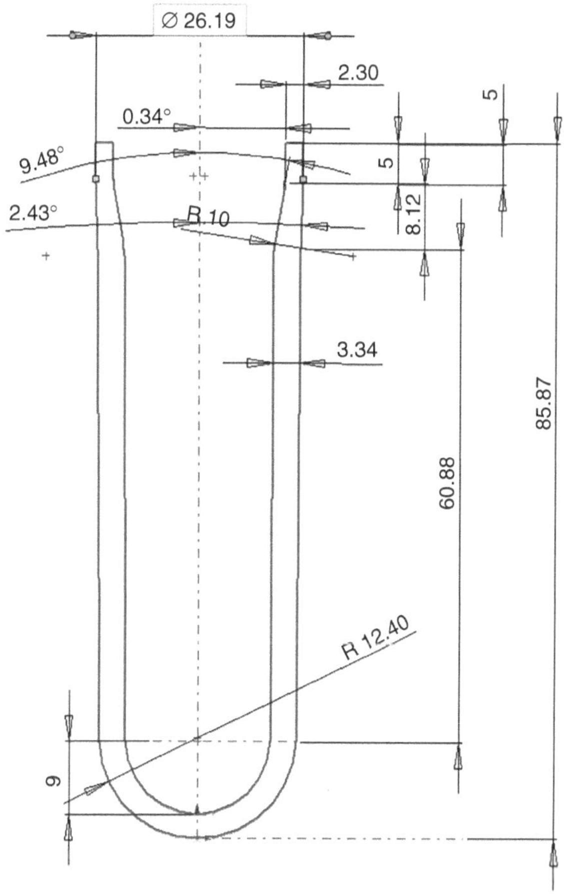

Figure 12.13 Preform for 1 L bottle step corrected.

be matched with the correct neck and a unit cavity developed. This additional development step is not only necessary to verify the weight and shape of the preform but also to make sure it can be blown into a suitable bottle. In the end, it is the bottle properties that count and all parties involved would do well to test any preform design before purchasing an expensive injection tool.

13 Auxiliary Equipment

Chapter Outline

13.1 Compressors

The compressor is the most important and most expensive piece of auxiliary equipment. Compressor selection is therefore of the utmost importance. Many plants run without any backup compressor because of these high costs and production comes to a halt whenever the compressor goes down.

To illustrate the high cost of compressor let us assume that the electricity cost is US$ 0.12 per kWh. In this case just 1 m^3/h of compressed air used over 8000 h will cost approximately US$ 117 (1 cfm would be around US$ 200).

Function of a Piston Compressor

Most compressors used in the reheat stretch blow molding (RSBM) process are of the reciprocating piston type with centrifugal types just starting to being developed. The former are referred to as "constant-volume/variable-pressure" machines. Volumes of air are taken into the confined compression chamber where a piston reduces the available volume thus increasing the pressure and heat as a by-product. A piston travels within a cylinder in four discrete steps (Fig. 13.1):

Stretch Blow Molding. http://dx.doi.org/10.1016/B978-0-323-46177-1.00013-5

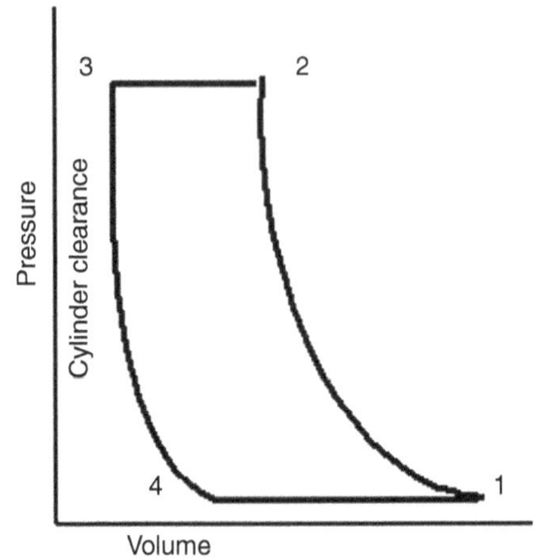

Figure 13.1 Simplified compressor cycle (valve losses not shown): 4–1, suction; 1–2, compression; 2–3, delivery; 3–4, reexpansion.

1. Suction: The piston moves away from the cylinder head, the suction valve opens and the cylinder fills with intake air.
2. Compression: The suction valve closes and the piston moves toward the cylinder head compressing and heating up the air within.
3. Delivery: The delivery valve opens and newly compressed air is pushed out of the cylinder.
4. Reexpansion: Clearances in the cylinder, as well as axial play in the piston rod, result in a small area inside the cylinder that cannot be used for pressure generation. Instead, during the suction stroke the compressed gases from this area reexpand. As it requires more energy to compress air in this space than is released during reexpansion, these clearances reduce the volumetric efficiency of the compressor and must be kept to a minimum.

During the compression stage the temperature and forces on linkages rise significantly while the amount of delivered air decreases. Therefore, compression to 40 bar takes place in three or four steps whereby each subsequent cylinder is smaller in diameter than the previous one. After each step, the air is cooled before it is further compressed. Once at 40 bar it runs through an after-cooler, reducing temperature a final time, and an air dryer that removes residual moisture. From there the air usually travels to an air receiver. Here air is stored and control logic allows the compressor to idle when the receiver is at full pressure.

Despite a suitably sized air dryer, moisture will still be present in the air and can make its way into machines and bottles. For this reason all pipes must have a slope of approximately 2.5 cm per 3 m (1 in. per 10 in.) of pipe. These pipes then end in vertical drops with automatic discharge valves that regularly exhaust accumulated moisture. Since water may run along the bottom of the pipes, connectors dropping down to machines should be connected to the upper half of the pipe diameter. Air receivers also collect water at the bottom that must be discharged as well on a regular basis.

Normally, a piston-type machine would be carefully lubricated to assure long life of its components. In blow molding for food applications this requires filtering out any and all oil particles that will become part of the air stream. While some compressor manufacturers have gone that route others offer oil-free compressors. While air from this type of compressor still needs to be filtered for particles that may have come loose from normal abrasion the number of particles is that much smaller and easier to filter out. Maintenance intervals may be shorter with oil-free compressors and should be carefully scheduled to avoid unplanned production stops.

Calculation of Air Requirements

There are several steps involved in calculating air demands for a blow machine, which are outlined here. Let us first look at all the parameters required for the calculation:

1. Blow machine output in bottles per hour (O).
2. Volume of largest container in cm^3 (V).
3. Lost volume of each cavity: This is the air space between the blow valves and the cavity. It has to be filled with each blow cycle but does not contribute to blowing the container. Due to the high cost of compressed air, this might be an important criterion for selecting a new machine.
4. Pressure in bar (P): Assume 35 bar for water, 40 bar for carbonated soft drinks bottles.
5. Safety factor (SF): Assume 10–20%.
6. Total air demand (TAD) will be the resulting value in cubic meter per hour.

The formula is:

$$\text{TAD} = \frac{O \times (V + \text{LV}) \times P \times (1 + \text{SF})}{1{,}000{,}000}$$

To convert to cfm, multiply this value by 1.7.

All compressors are rated in either normalized cubic meters per hour (Nm^3/h) or standard cubic feet per minute (Scfm). This calculation assumes the compressor to be at sea level and the intake air to be at 20°C (68°F) and 36% humidity. These three factors have a significant impact on air delivery capacity:

1. Altitude of plant: Intake air pressure decreases with altitude leading to higher demands on the compressor. Altitude charts are available on the Internet.
2. Temperature of plant: Higher temperatures lead to thinner air. Assume maximum temperature during summer months. This is independent of whether machines have dehumidification devices.
3. Humidity in plant: Higher moisture content reduces capacity. Assume maximum summer humidity.

Tables 13.1 and 13.2 lists the relevant correction factors. The value TAD has to be divided subsequently by the factors to arrive at the site air demand (SAD):

$$\text{SAD} = \frac{\text{TAD}}{(\text{Cor1} \times \text{Cor2} \times \text{Cor3})}$$

Compressor Selection

There are a variety of compressor designs within the reciprocating piston compressor category. The main distinction is the way in which the cylinders are mounted relative to each other. V-shape, L-shape, horizontally opposed, as well as in-line solutions are all common and each has its own advantages and disadvantages. What is most important to the end-user is the ratio of power consumption to volume of delivered air. This ultimate performance gauge will reveal all efficiencies throughout the system and can distinguish systems for end-users who are often at a loss in determining the best compressor solution (Fig. 13.2).

Other criteria to consider are as follows:

- manufacturer's recommended maintenance schedule and
- availability of spare parts and service.

Table 13.1 Relevant Correction Factors

Correction Factor/Altitude			Correction Factor/ Temperature	
Altitude (m)	Atmospheric Pressure (bar) abs.	Correction Factor	Temperature (°C)	Correction Factor
0	1.013	1	0	1
500	0.952	0.94	1	1.00
750	0.923	0.91	14	0.95
1000	0.894	0.88	15	0.95
1100	0.883	0.87	16	0.94
1200	0.872	0.86	17	0.94
1300	0.861	0.85	18	0.94
1400	0.850	0.84	19	0.93
1500	0.840	0.83	20	0.93
1600	0.829	0.82	22	0.93
1700	0.819	0.81	24	0.92
1800	0.808	0.80	26	0.91
1900	0.798	0.79	28	0.91
2000	0.788	0.78	30	0.90
2100	0.778	0.77	32	0.90
2200	0.768	0.76	34	0.89
2300	0.759	0.75	36	0.88
2400	0.749	0.74	38	0.88
2500	0.739	0.73	40	0.87
2600	0.730	0.72	42	0.87
2700	0.721	0.71	44	0.86
2800	0.711	0.70	46	0.86
2900	0.702	0.69	48	0.85
3000	0.693	0.68	50	0.85

All data provided by AF USA.

13.2 Chillers

Although the chiller does not play an important role in blow molding as it does in injection molding, proper lay-out and capacity is paramount for a smooth plant operation (Fig. 13.3).

All chillers use a refrigerant that changes states through the system. The chiller's compressor forces it into a high-pressure, high-temperature

Table 13.2 **Correction Factor/Relative Humidity**

Correction Factor/Relative Humidity						
	Correction Factor at Temperature (°C)					
RH (%)	20	25	30	35	40	45
50	0.9886	0.9845	0.9792	0.9723	0.9637	0.9528
55	0.9875	0.9829	0.9771	0.9696	0.96	0.948
60	0.9863	0.9814	0.975	0.9668	0.9564	0.9433
65	0.9851	0.9798	0.9729	0.964	0.9527	0.9386
70	0.984	0.9782	0.9708	0.9612	0.9491	0.9338
75	0.9828	0.9767	0.9687	0.9584	0.9454	0.9291
80	0.9817	0.9751	0.9666	0.9556	0.9418	0.9243
85	0.9805	0.9735	0.9645	0.9529	0.9381	0.9196
90	0.9794	0.972	0.9624	0.9501	0.9345	0.9149
95	0.9782	0.9704	0.9603	0.9473	0.9308	0.9101
100	0.9771	0.9688	0.9582	0.9445	0.9272	0.9054

All data provided by AF USA.

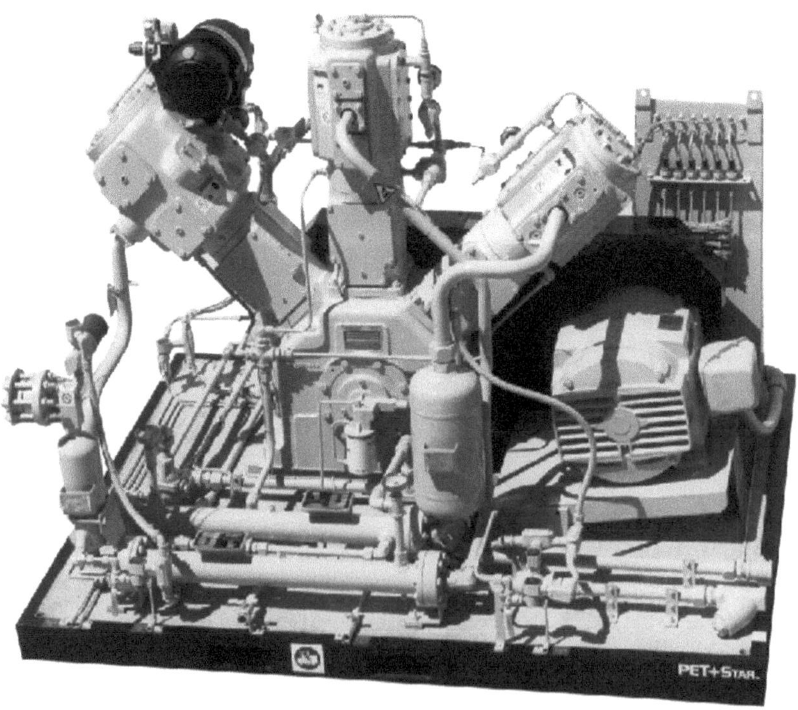

Figure 13.2 Reciprocating piston compressors are most common in the PET industry. Picture courtesy of Ingersoll-Rand.

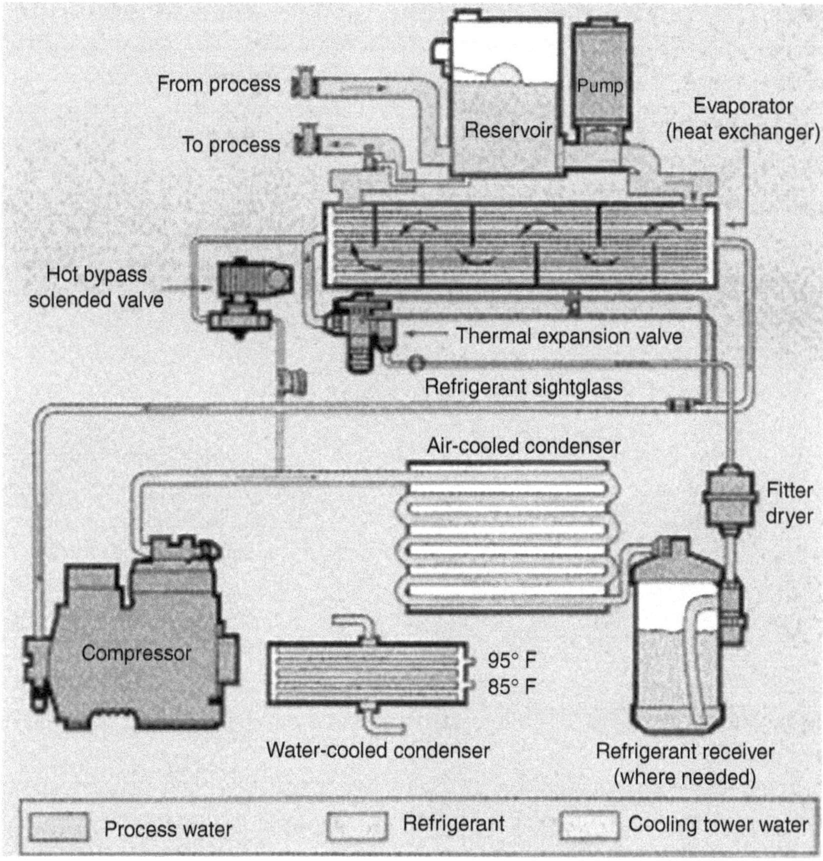

Figure 13.3 Chiller refrigeration cycle. Diagram courtesy of Schaub Chiller Service.

gas. In this state it travels to a condenser that may be air- or water-cooled. The condenser changes it to a warm liquid. From there it may travel to a receiver (that is not always required). The next stop is the thermal expansion valve that changes the refrigerant into a low-pressure, cold-saturated gas and meters it into the evaporator. Here the gas changes into a cool and dry gas which can take on heat from the process water.

Only process cooling requires low temperatures of 7–13°C (45–55°F). Other cooling requirements can be satisfied with water at 23–29°C (75–85°F). It can therefore be advantageous to separate these requirements and have two cooling systems. The system for the former would be a chiller, whereas the latter can be serviced with an evaporation device such as a water tower. The water tower would then also cool the condenser of the chiller unit. This is a *water-cooled* system. Alternatively, fans could do all cooling and this *air-cooled* chiller could be made portable and could

Table 13.3 Advantages and Disadvantages of Water- and Air-Cooled Systems

Item	Water-Cooled System	Air-Cooled System
Investment/output	Good	Fair
Energy/output	Good	Fair
Make-up water required	Yes, because of cooling tower evaporation	No, only occasionally
Maintenance	Fair	Good
Portability	No, centralized system	Yes
Water treatment internal	Good	Good
Water treatment external	Fair; evaporation causes concentration of mineral content	None
Use for heating in cold months	No	Yes. Air stream from fans may be diverted into plant for part of the year
Space requirements	Good	Fair

sit beside the machine. Table 13.3 lists the advantages and disadvantages of both the systems.

In short, water-cooled systems are less expensive and more energy-efficient whereas air-cooled systems require less maintenance, are portable, and can be used for heating the plant (Fig. 13.4).

Water treatment is important for both systems but more so for the water-cooled option. Since evaporation in the water tower buildup of minerals and other contaminants requires careful dosing of appropriate chemicals and frequent water checks. Untreated water will lead to corrosion in pipes and machines as well as to scale and algae buildup. These deposits stick to pipe walls and form an insulating barrier, thus reducing system efficiency. This in turn forces motors and pumps to work harder leading to premature breakdown and costly production interruptions. Users are well advised to have site water analyzed and a water treatment formula devised that will keep deposits in check. Automatic systems are available that can be programmed to release treatment agents just in time.

Antifreeze

The use of antifreeze in cooling water deserves a special note. Antifreeze has two functions in a water system.

Figure 13.4 Central water chilling system. Photo courtesy of AEC, Inc.

As the name implies it reduces the freezing point of water. Water temperature at the contact point in the evaporator will be several degrees lower than at the exit point of the system. Hence it may reach its freezing point even though the chiller is adjusted to a temperature well above it. The required antifreeze concentration will depend on the chosen water temperature. Antifreeze also has an anticorrosive effect. However, using antifreeze for corrosion protection requires concentrations of about 25%, which is the reason why this concentration level is recommended in many manuals. Unfortunately the ability of antifreeze to absorb heat from mold parts is inferior to that of water and the efficiency of the chiller system can be compromised. Furthermore, antifreeze is more viscous than water and so increases pump loads.

It is therefore preferable to adjust the antifreeze concentration only with respect to lowering the freezing point as necessary and add corrosion protection by other means with regular water treatment (Table 13.4).

Necessary antifreeze concentration depends on cooling water temperature.

Antifreeze cannot be used in a cooling tower because the performance of a cooling tower is based on evaporation. Adding antifreeze continuously

Table 13.4

Water Temperature (°C)	Water Temperature (°F)	Glycol Required (%)
10	50	1
9	48	3
8	46	5
7	45	7

into cooling tower water would be expensive and is illegal in many jurisdictions as the evaporating antifreeze is an environmental hazard. Freeze prevention is accomplished instead by using a thermostat to cycle the tower fan and the tower recirculation pump on and off based on a temperature set point of between 21 and 29°C (70 and 85°F).

Antifreeze comes in two formulations: ethylene–glycol and polypropylene–glycol. The former is a controlled substance that may not be released into the municipal water system. The latter is environmentally friendly and should be used wherever available.

Flow in Cooling Systems

Process heat flows to the chilled water in three different ways:

1. polyethylene terephthalate (PET) to mold metal
2. mold metal to coolant in cooling channel
3. from outside of the mold (shop air) to the non-PET side of the cooling channel (so-called “mold sweating”)

The first step is controlled by the heat conductivity of the mold material (Chapter 6). The second step depends on how much contact cold cooling agent makes with the PET side of the cooling channel. This in turn is controlled by the way the cooling agent flows and its distance from the hot PET.

We distinguish two flow modes in liquids: laminar and turbulent. Laminar flow is characterized by the highest speed in the center of the channel where there is little friction decreasing to zero at the wall of the cooling channel where friction is highest (Fig. 13.5).

Heat exchange between cooling agent and mold metal is greatly jeopardized by the slow speed in the contact area. Systems with laminar flow show a large increase in coolant temperature through the system.

Turbulent flow on the other hand allows frequent contact between coolant and mold metal. Due to the constant mixing of the coolant its temperature does not spike up. Turbulent flow is not so much dependent on a high

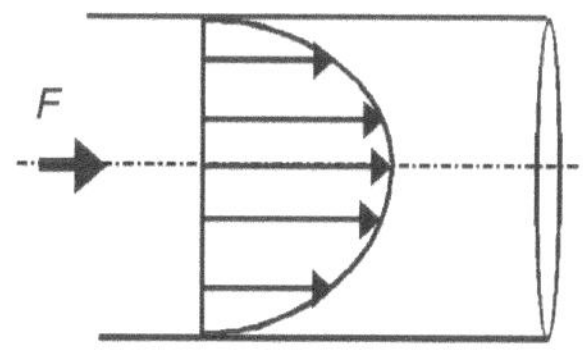

Figure 13.5 Velocity profile in laminar flow. *Longer arrows* indicate higher speeds.

inlet pressure but rather on the pressure difference between the inlet and outlet. Pressure drop of 5 bar (70 psi) is sufficient to achieve this effect. Sizing return water line diameters to 25 mm (1 in.) bigger than supply lines often reduces back-pressure that might be present in the system. A good sign of turbulent flow is a temperature difference of 1.5°C (3°F) between the inlet and outlet points. (Measurements must be taken by thermocouples inside the coolant flow, not on the pipes.) Increasing pump size in the central location or at each machine (booster pumps) is the most likely remedy for any deficiencies.

If exact calculations are required, the so-called "Reynolds number" must be determined. This dimensionless number takes into account the viscosity of the fluid, the mass velocity, and the length of the pipes. A good resource for this can be found at: http://www.processassociates.com/process/dimen/dn_rey.htm. Reynolds numbers above 2000 indicate turbulent flow.

Calculation of Chilled Water Requirements

There are a number of units used to describe chiller capacity, the most important ones are listed in Table 13.5.

There are essentially three types of cooling requirements:

1. Cooling PET bottles from blowing to demolding temperature.
2. Cooling of hydraulic oil.
3. Cooling of auxiliaries such as compressors.

Table 13.5 Units Used to Describe Chiller Capacity

Unit	Abbreviation	Conversion
Kilogram–calorie	kcal	1 kcal = 4.19 kJ (kilo-Joule)
British thermal unit	Btu	1 kcal = 3.97 Btu
Frigorie	Frig	1 Frig = 3.97 Btu
Tons	Tons	1 ton = 12,000 Btu chiller
		1 ton = 15,000 Btu water tower

Process Cooling

Here are the parameters needed to complete the calculation:

1. Blow machine output in bottles per hour (*O*).
2. Heaviest bottle to be produced in grams (*W*).
3. Delta *T* (DT) of bottle: Assume 70°C (110–40°C) (126°F).
4. Specific heat (SH) of PET: At 100°C this value is 1.4 kJ/(kg × K) [0.33 Btu/(lb. ×°F)].
5. SF: Assume 10–20%.
6. Process pump horsepower (PHP).
7. Duty cycle (DC) of process pump (assume 50%).
8. Result is process heat (*Q*) per hour.

$$Q = O \times W \times \text{DT (in °C)} \times \text{SH (in kJ)} \times \text{SF} / 1000 + \text{PHP} \times \text{DC} \times 1581 \text{ (in kJ)}$$
$$Q = O \times W \times \text{DT (in °F)} \times \text{SH (in Btu)} \times \text{SF} / 454 + \text{PHP} \times \text{DC} \times 1500 \text{ (in Btu)}$$

This indicates the amount of cooling the chiller must provide.

Hydraulic Oil Cooling

If the blow molding machine uses hydraulic pumps oil heats up during machine operation and must be cooled to a temperature not exceeding 55°C (131°F). This can be accomplished by either chilled or tower water.

Horsepower in this calculation refers to the horsepower of the pump motor on the blow molding machine.

If chilled water is used the calculation is:

$$\text{Tonnage} = \frac{\text{hp} \times \text{DC} \times 2{,}684 \text{ kJ } (2{,}547 \text{ Btu})}{12{,}648 \text{ kJ } (12{,}000 \text{ Btu})}$$

If tower water is used the formula is:

$$\text{Tonnage} = \frac{\text{hp} \times \text{DC} \times 2{,}684 \text{ kJ } (2{,}547 \text{ Btu})}{15{,}810 \text{ kJ } (15{,}000 \text{ Btu})}$$

Cooling of Auxiliaries

Other cooling requirements come from machines such as compressors, material dryers, as well as water-cooled heat shields and reflectors if applicable. All equipment specifications identify the amount of water flow needed and the difference between water inlet and outlet temperatures. To convert this information into chiller tonnage use the formula:

$$\text{Tonnage} = \text{Flow in m}^3/\text{h} \times \text{DT in °C} \times 4{,}078/12{,}000 \text{ or } 15{,}000 \text{ Btu}$$
$$\text{Tonnage} = \text{Flow in gpm} \times \text{DT in °F} \times 500/12{,}000 \text{ or } 15{,}000 \text{ Btu}$$

Pump Sizing

These approximations can be used to calculate the proper size of pump for your system.

Chiller:	30 L/min (8 gpm) for each ton of chilling capacity
Water tower:	13.2 L/min (3 gpm) for each ton of chilling capacity

Using the flow rate obtained in this way the nominal brake horsepower can be calculated, entering 6–7 bar (87–101 psi) as delivery pressure. This value must then be correlated to actual pump curves resulting in a somewhat higher pump horsepower actually required. Both chiller and water tower should have standby pumps. These may be alternately used on a weekly basis and act as backup should one of them go down. It is a relatively inexpensive investment and will avoid production stops due to pump failure.

Most chillers are rated on the basis of 10°C (50°F) water supply temperature. For every 0.6°C (1°F) drop in temperature from this figure, the chiller loses about 2% of its capacity. Running water below this threshold is usually not required in RSBM.

13.3 Conveyors and Bottle Storage Devices

Conveyors

Most conveyors used in high-speed applications are air conveyors. Bottles slide by their neck support rings and air blowers spaced in intervals of 6–15 m (19–50 in.) propel them at speeds of up to 1000 bottles/min. Blowers are configured for high volume and low speed, gently blowing at the bottle shoulders and sidewalls. Stand-up or belt conveyors may also be used but unless they are correctly adapted may not be suitable to convey today's lightweight bottles. The conveying air should be filtered and the conveyor covered to avoid contamination. Some companies offer

Figure 13.6 Palletizer/depalletizer systems greatly increase buffer time and still keep bottle orientation. Picture courtesy of Goldco Industries, Inc.

quick-change systems for different neck sizes thus reducing changeover times. These conveyors may be totally enclosed for aseptic filling lines.

Palletizers/Depalletizers

Palletizer and depalletizer systems organize bottles in programmable patterns and assemble them onto skids. There are a variety of levels of automation available. Most systems automatically dispense the wooden or plastic pallets as well as cardboard sheets that are inserted between each layer of bottles, and deliver these skids onto a conveyor belt. From there they may be taken away by fork lifts and stored. Fully automatic systems are designed for a certain amount of bottles that translate into a certain buffer time depending on throughput and bottle size. These systems then automatically palletize and depalletize bottles depending on supply and demand parameters monitored by line sensors. Suitable gates move to divert bottles in and out of the system (Fig. 13.6).

Silos

Silos are another means of storing bottles. They can be either directly in-line, where all bottles are conveyed to them, or the silo may be used as a temporary buffer. In either case only careful design of chutes and baffles

Figure 13.7 Silos provide the longest buffer time and may allow the blow molding machine to run longer production hours to keep pace with faster downstream equipment that runs less hours/day or week. Picture courtesy of Riondé S.A.

avoids bottle damage and jams. Trials should be undertaken to determine suitability of bottles, especially when they are lightweighted (Fig. 13.7).

Silos use inclined bulk conveyors to transport bottles to the top entry point. These should be covered and pressurized by blowers to keep dust and other contamination out. Filtered cool air may also be used to cool bottles as they are moving, making them less susceptible to damage.

Table 13.6 Relationship Between Bottle Size and Required Silo Volume

Bottle Size (L)	Bottles (m^{-3})	Bottles ($ft.^{-3}$)
0.1	4200	119
0.25	1550	44
0.33	1130	32
0.5	780	22
1	400	11
1.25	350	10
1.5	320	9
2	220	6
3	120	3

Silo storage capacity can be calculated by multiplying upstream machine throughput and the anticipated maximum buffer time that will be required. The resultant number of bottles will need a certain silo volume depending on bottle size. Table 13.6 lists the relationship between bottle size and required silo volume.

Approximate bulk density of various bottle sizes. Table 13.6 may not be accurate for every bottle geometry.

The volume calculated in this way must then be translated into silo dimensions suitable for the given factory environment. It should be noted that silos will occupy a significant higher volume than the storage space they provide.

Unscramblers

There are a variety of different designs in use to take scrambled bottles and orient them onto a conveyor. Most common are bowls of various diameters that contain rotating disks. Centrifugal forces take bottles to the rim of the bowl where a series of levers and guides bring about orientation (Fig. 13.8).

Unscrambling PET bottles has become more of a challenge with the advent of lightweight bottles and not every bottle design will withstand unscrambling without damage. When in doubt, end-users should submit sample bottles in sufficient quantities for a test run on the prospective equipment before making a purchase decision. Bottles may then be evaluated with respect to scuffing and other damage. It should also be noted whether the unscrambler is able to remove damaged bottles automatically and how many jams (if any) may happen during a few hours of running.

Figure 13.8 Most common unscramblers feature rotating disks. Picture courtesy of Lanfranchi srl.

14 Training of Operators

Chapter Outline

14.1 Current State of Training

Many professionals decry the current state of training in blow molding plants as unsatisfactory. This lack of knowledge presents itself in both captive and noncaptive operations. However, it is more evident in the latter because of the frequent changeovers and the greater number of different types of bottles blown. Visible problems in a typical blow molding plant can be seen in the following:

- Pack-out rates well below the achievable 95%.
- Stretch rod adjustments causing wall thickness variations.
- Timing settings leading to inferior bottle properties.
- Injection profiles that cause process variations in single-stage machines.
- Insufficient airflow in ovens of reheat stretch blow machines causing convective heating of preforms.

The cost to the industry of these and other problems are significant. This is most evident in the cost of wasted material. As the output per cavity

Stretch Blow Molding. http://dx.doi.org/10.1016/B978-0-323-46177-1.00014-7

and the number of cavities of both injection and stretch blow machines reach ever new heights, even short periods of running nonsellable products amount to large number of scrap preforms or bottles that are costly to sort or grind. Machine damage, personal injuries, loss of customer confidence in passing deadlines can be improved by well-trained operators knowledgeable in the process and its effects on bottle quality. Personnel with a high level of job satisfaction will consistently maintain quality production.

14.2 Obstacles to Training Programs

Many operators in the industry have never been in any formal training program and some of them are excellent in their job skills. This raises the question of whether training is the method of choice in an industry where some black art seems necessary to get machines to work. One should consider however, that many of these present-day blow molding experts learned their trade in the 1960s and 1970s. During that period, machines, time, and material were far less costly and the drive to on time and just-in-time delivery was still a far cry away. In today's fast-paced environment, it is much harder to allow people to learn from their mistakes on the job. Companies have to be able to rapidly adjust to changing market conditions and respond flexibly to customer demands. Nevertheless, plant managers have concerns making them reluctant to start training programs. The four most prevalent are:

1. Training is too expensive: It is often difficult to find money in training budgets already strained by safety and ISO issues. However, operator training can be the most profitable investment a company can make.
2. Trained personnel leaving for better paying positions elsewhere: A plastics industry with a regulatory framework and a set standard for all operators would alleviate present discrepancies between personnel in different plants.
3. Training is ineffective.

This last point warrants careful consideration. Many managers observe personnel trained in formal plastics programs (schooled or in-house trained), are not adequately prepared to actually control the machines in their plants. Training often does not translate into better quality production. The present notion is a good operator is only created by hands-on machine experience. There are difficulties in teaching blow molding techniques that training programs need to overcome.

14.3 Rethinking Abstract Concepts

There is certain unpredictability in the blow molding process. Preforms have encapsulated different stress levels from their heat history and pressure environment. They are then exposed to varying factory temperatures and humidity conditions. Machine controls and wear conditions differ slightly from machine to machine. All this affects the process and requires an almost intangible skill on the part of the operator that needs to work through complex concepts and arrive at a solution for a particular problem. Most operators have a vast arsenal of visual and tactile experiences they use every day to tackle machine and process problems. Trainers need to facilitate an enhancement of these experiences with the abstract concepts of polymer science. The substitution of a real-life experience or measurement to a letter or number needs to be helped by careful wording that is close to practical knowledge. Operators are often intimidated by the presentation of unfamiliar materials. When this happens, the learning experience can be seriously impeded. Therefore, trainers need to connect with what the operators know already and show the underlying principles at work.

14.4 Language Structure

The wording and grammatical structure of most training material is rooted in the minds of the engineers that conceived them. Their language is not spoken on the floors of blow molding plants. Language used in written material presented using the trainees' terms and structure will maximize comprehension levels. Engineering terms are very concise expressions that usually have a lot of precise meaning. However, short sentences are better to convey the meaning of a single term. Here are some examples:

"Amorphous state": The state of the plastic when it is without orientation or structure, imagined as a bowl of spaghetti.

"Intrinsic viscosity": The number by which we measure how long the individual strains of plastic are and how much they resist flow.

14.5 Converting Formulas Into Common English Sentences

Formulas can be overwhelming to people that never use them in their everyday life.

Take this simple formula, for example:

$$O = \frac{3600}{\mathrm{CT} \times \mathrm{CA}} \tag{14.1}$$

where O, output/hour; CT, cycle time; CA, cavitation.

While this is a perfectly fine notation, the translation below will be better understood by many: to get the hourly output of a bottle machine, divide 3600 by the cycle time, and then multiply by the number of cavities.

14.6 Substituting Uncommon Arithmetical Operators

These include: “<,” “>,” “&,” “‰,” and so on. Replace these with “smaller than,” “greater than,” “and,” and “per thousand.”

14.7 Presentation Style

In order for the teacher/student classroom environment to become a rewarding experience, trainers need to adopt a suitable presentation style that keeps trainees focused and interested. The trainer should stand, walk, and use arms and facial expressions to make the presentation livelier. The initial awkwardness of the situation can be overcome by telling some interesting stories with funny components out of the trainer’s experience in the field. Asking questions is however, the best tool to get trainees thinking and check their comprehension levels. It is therefore a good idea to ask questions before getting into a topic. These will show how much knowledge there is already, and help the trainer to adopt the material to the particular needs of this group of trainees. Questions asked about the subject already discussed, will show the trainer where the gaps in the knowledge of the subject are and what he/she can do to overcome them.

14.8 Translating Graphs into Common-Day Language

The interpretation of graphs is a skill learned in college and university courses. Difficulties can arise when trainers use them to demonstrate a point. Many operators have trouble correlating a x- and y-axis with any useful values. Consider the graph of equilibration time versus preform

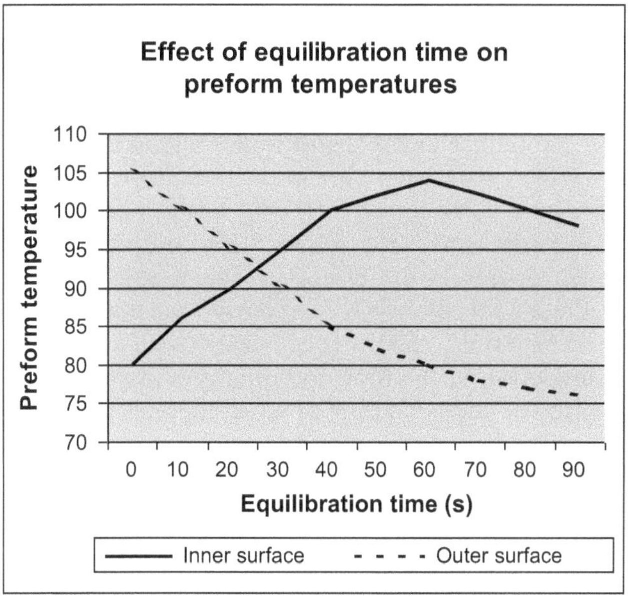

Figure 14.1 All graphs used in training must be carefully explained.

temperatures (Fig. 14.1). While the graph can be used during the training, the following interpretation needs to be attached.

"The time between the end of reheating to the start of stretch blowing is called equilibration time, because the preform temperatures adjust once the preform is no longer heated up. From the graph, we can see that at the end of heating, the outer surface temperature is higher than the inner surface temperature because the radiation from the lamps hit the outer surface first. As time goes by, the outer surface cools down because it is cooled down by the air around it. The inner surface however, heats up because it is not cooled down and heat from the outer surface penetrates to the inside wall. After 30 s, the two temperatures are about the same. After 65 s, the inner surface is now 25°C (45°F) higher than the outer surface."

In US plants, all temperatures would be displayed in Fahrenheit only.

14.9 Choosing Easy-to-Understand Drawings

Most technical drawings contain too much detail and are hard to comprehend for people that have never had a formal education to read them. Three dimensional drawings of the most important components if necessary layered to show more detail, are comprehended with much better success.

14.10 Computer Simulations

There is no publicly available training material for stretch blow molding at this point in time while a number of materials can be purchased for injection molding and extrusion blow molding. Some companies as well as outside consultants have created their own media but this is only available with these companies. One way out of this situation may be to use computer simulation to teach key subjects. It can be quite costly to teach operators the effects of machine settings and design features on the actual, often very expensive machinery. The machine downtime and material scrap produced can exceed any training budget. Computer simulation of the blow molding process offers a convenient way to increase the understanding of what happens inside a blow mold (Fig. 14.2).

In this example, the preform is marked with a dot in the center position of the first infrared lamp. The mark can be traced to the bottle and the effect of heating this area up to a lesser or higher degree on bottle wall thickness can be studied. Operators learn important concepts out of the training without tying up a $2 MM machine.

14.11 Comprehensive Coverage

It is often amazing to see how ingenious operators can be in keeping machines running although their solutions are not based on a solid knowledge of the underlying science. During the training, the trainer may find out that operators lack basic principles but have managed to work around this lack by finding alternatives to a proper solution. Therefore, all angles of the material should be covered and by evaluating the answers to specific questions, the training can then be directed in the areas where operators need help the most.

14.12 Trainees Involvement

Through the whole training session, it is important to get the operators involved. The best way is to pose questions that are then answered by the material taught. Here are some examples:

"How is it possible that a white PET pellet can produce a crystal clear preform?"

"What happens when the high-pressure air comes on before the stretch is all the way extended?"

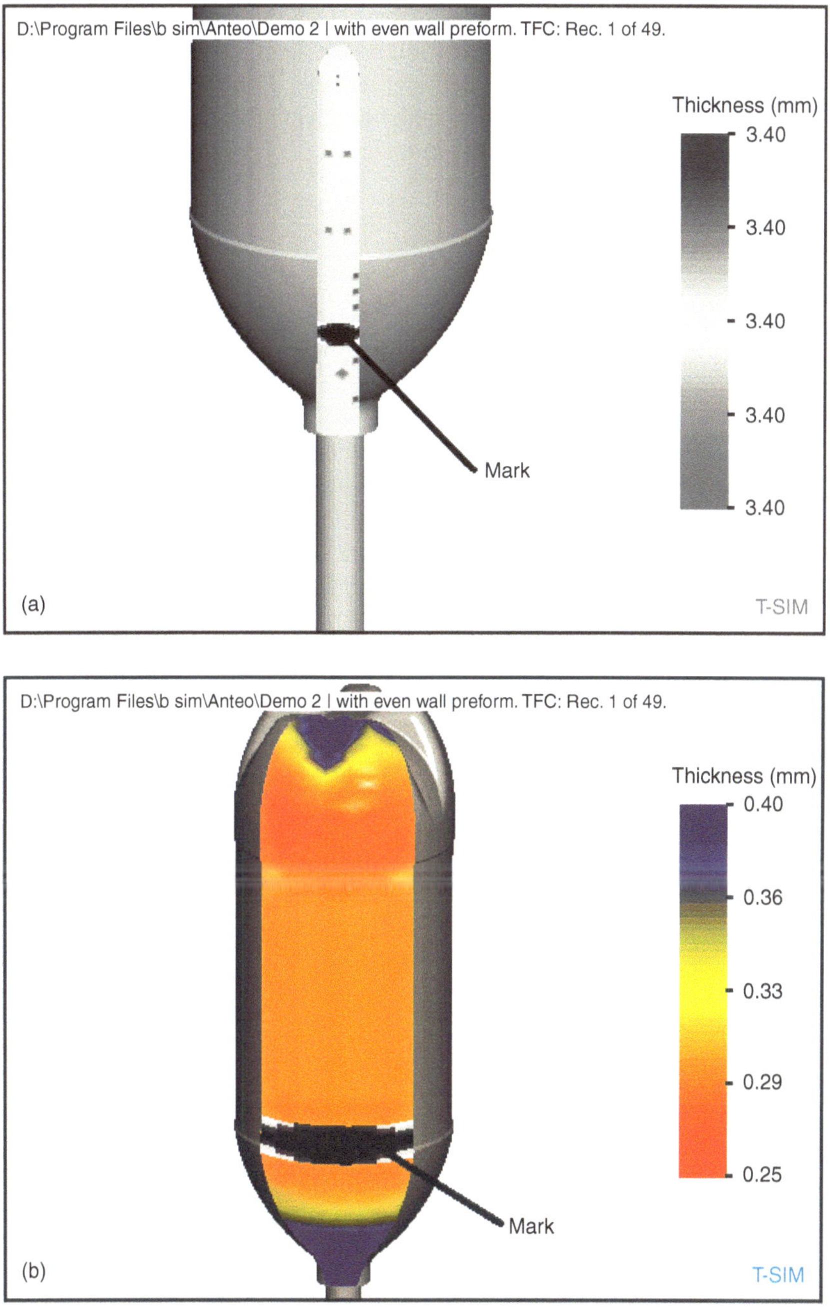

Figure 14.2 Preform (a) and bottle (b) with mark. Computer simulation can explain key concepts. Diagrams produced with B-Sim software.

"What is the last action to take place before the single-stage injection stretch blow machine starts a new cycle?"

At various times during the training, senses other than hearing need to be involved. Having samples of plastic bottles or tool components at hand offer an easy connection to tactile knowledge.

14.13 Assessments

Many operators are reluctant to take tests. There is a fear that poor test results will lead to fewer promotion opportunities. However, it is not possible to judge the success of a training session without this feedback. Tests should be used to evaluate operator skills before the training so the program can be structured to their needs. Additionally, the operators should be given a choice of topics that they would like the training to concentrate on. This will help with acceptance of the training process. At the end of the training, a multiple-choice test should be structured to cover the material taught and to give helpful clues about essential topics. The corrected test is then returned to the operators for them to use as reference material. Many operators use these corrected tests more than other training material when they are faced with a particular problem on the machine.

In-Plant or Off-Plant Training

Both approaches have their own distinct advantages and disadvantages.

In-plant training can be combined with hands-on experience of the material taught in a familiar environment. This is by far, the best way to reinforce what was taught. The disadvantage is that it is difficult to create a calm atmosphere and people are often being called out of the training room to attend machinery. Another obstacle is the noise level in a typical production environment. It is virtually impossible to talk to more than two people at a time with the result that the most eager students crowd the trainer while less enthusiastic ones get bored. Planning the trip to the plant in the classroom can help but often circumstances are somewhat different than what was anticipated and the plan has to be adjusted. A good solution is to use teacher/student headsets with special noise-canceling devices.

Off-plant training does not have these problems but often lacks the hands-on component that is so important. Even when machines are available, these often differ quite a bit from the machines the operators are used to and are not as successful in reinforcing training material.

If an undisturbed environment in the plant can be guaranteed where no trainee will be called away, in-house training is the best way to go. It is also more economical. If not, it might be better having the training in a separate location.

14.14 Conclusions

There are a number of obstacles to a successful training experience. Trainers can overcome these obstacles by tailoring the material to the specific requirements of the trainees. They can use any media as long as it is clear and easy to understand. Simple tables and instructions are preferable to convoluted high-tech presentations. Difficult concepts can be broken down into the main components and from there workable, practical solutions can be developed.

Training blow molding operators is a challenge. However, blow molding is a people business. This is because skilled operators cannot be replaced easily and machines do not run on their own (with some exceptions, see Chapter 10, Section 10.7). As machines become more expensive, every hour they are not running optimally, costs the company more and more money. Therefore, companies are well-advised to foster a group of dedicated professionals that can make this happen. Developing optimal training methods that significantly improve the skills of floor personnel will increase the bottom line and open new applications for blow molding.

Index

Papier | Fördert gute Waldnutzung
FSC® C083411

Kolibri 360 GmbH
Ferdinand-Jühlke-Straße 7
99095 Erfurt, Deutschland
produktsicherheit@kolibri360.de